Molecular Basis of Human Cancer

NATO ASI Series

Advanced Science Institutes Series

A series presenting the results of activities sponsored by the NATO Science Committee, which aims at the dissemination of advanced scientific and technological knowledge, with a view to strengthening links between scientific communities.

The series is published by an international board of publishers in conjunction with the NATO Scientific Affairs Division

A	**Life Sciences**	Plenum Publishing Corporation
B	**Physics**	New York and London
C	**Mathematical and Physical Sciences**	Kluwer Academic Publishers
D	**Behavioral and Social Sciences**	Dordrecht, Boston, and London
E	**Applied Sciences**	
F	**Computer and Systems Sciences**	Springer-Verlag
G	**Ecological Sciences**	Berlin, Heidelberg, New York, London,
H	**Cell Biology**	Paris, Tokyo, Hong Kong, and Barcelona
I	**Global Environmental Change**	

Recent Volumes in this Series

Volume 205—Developmental Patterning of the Vertebrate Limb
edited by J. Richard Hinchliffe, Juan M. Hurle, and Dennis Summerbell

Volume 206—Alcoholism: A Molecular Perspective
edited by T. Norman Palmer

Volume 207—Bioorganic Chemistry in Healthcare and Technology
edited by Upendra K. Pandit and Frank C. Alderweireldt

Volume 208—Vascular Endothelium: Physiological Basis of Clinical Problems
edited by John D. Catravas, Allan D. Callow, C. Norman Gillis, and Una S. Ryan

Volume 209—Molecular Basis of Human Cancer
edited by Claudio Nicolini

Volume 210—Woody Plant Biotechnology
edited by M. R. Ahuja

Volume 211—Biophysics of Photoreceptors and Photomovements in Microorganisms
edited by F. Lenci, F. Ghetti, G. Colombetti, D.-P. Häder, and Pill-Soon Song

Series A: Life Sciences

Molecular Basis of Human Cancer

Edited by

Claudio Nicolini

University of Genoa
Genoa, Italy

Springer Science+Business Media, LLC

Proceedings of a NATO Advanced Study Institute/Eleventh Course of the International School of Pure and Applied Biostructure on the Molecular Basis of Human Cancer, held May 21–June 1, 1990, in Erice, Sicily, Italy

Library of Congress Cataloging-in-Publication Data

International School of Pure and Applied Biostructure (11th : 1990 : Erice, Italy)
Molecular basis of human cancer / edited bv Claudio Nicolini.
p. cm. -- (NATO ASI series. Series A, Life sciences ; v. 209)
"Proceedings of a NATO Advanced Study Institute/Eleventh Course of the International School of Pure and Applied Biostructure on the molecular basis of human cancer, held May 21-June 1, 1990, in Erice, Sicily, Italy"--T.p. verso.
Published in cooperation with NATO Scientific Affairs Division.
Includes bibliographical references and index.

1. Cancer--Molecular aspects--Congresses. I. Nicolini, Claudio A. II. North Atlantic Treaty Organization. Scientific Affairs Division. III. Title. IV. Series.
[DNLM: 1. Carciogens--congresses. 2. Molecular Biology--congresses. 3. Neoplasms--congresses. QZ 200 I60943m 1990]
RC269.I542 1990
616.99'407--dc20
DNLM/DLC
for Library of Congress 91-24230
CIP

ISBN 978-1-4899-2565-7 ISBN 978-1-4899-2563-3 (eBook)
DOI 10.1007/978-1-4899-2563-3

Originally published by Plenum Press, New York in 1991
Softcover reprint of the hardcover 1st edition 1991

PREFACE

During May 21-June 1 1990, the eleventh course of the International School of Pure and Applied Biostructure, a NATO Advanced Study Institute, was held at the Ettore Majorana Center for Scientific Culture in Erice, Italy, co-sponsored by the Italian Ministry of Universities and of Scientific and Technological Research, the North Atlantic Treaty Organization, the Italian National Research Council, the Sicilian Regional Government and Technobiochip. The subject of the course was "Molecular Basis of Human Cancer" with participants selected worldwide from 15 different countries.

The purpose of the course was to address, in a tutorial and structural fashion, the molecular basis of human cancer, including the mechanism of signal transduction in mammalian cells, the genetic mechanism of malignant transformation in man, growth factors, hormone receptors, cell membrane and cytoskeleton, and DNA high order structure. The course had this as its major objective and the resulting book reflects it. The participants were exposed to a critical evaluation of current knowledge about cancer and to some of the key problems that remain as stumbling blocks to our eventual understanding of this important biological and medical problem. Through the media of formal and informal lectures, workshops, symposia and informal discussions, a select group of interested young and senior scientists were acquainted with many of the aspects of human cancer.

This book is the result of this Advanced Study Institute and is the twelfth of a series, which began with "Chromatin Structure and Function" (1979, Vol. A21 & B) and continued with "Cell Growth" (1982, Vol. A38), "Chemical Carcinogenesis" (1982, Vol. A52), "Interactions between Electromagnetic Fields and Cells" (1985, Vol. A97), "Modeling and Analysis in Biomedicine" (1984, W.S.P.), "Structure and Function of the Genetic Apparatus" (1985, Vol. A98), "NMR in the Life Sciences" (1986, Vol. A107), "Cell Biophysics" (1987), "Towards the Biochip" (1988, W.S.P.), "Protein Structure and Engineering" (1989, Vol. A183), and Structure and Dynamics of Biopolymers" (1986, Vol. E133, Martinoff), edited or occasionally coedited by myself as Director of the International School of Pure and Applied Biostructure and published mostly by Plenum within the NATO ASI Life Science Series and partly by other publishers. This book aims to present a structured and interdisciplinary view of current knowledge on the possible molecular and cellular mechanisms leading to human cancer initiation and promotion.

I wish to express my gratitude to Carlo Croce for his active coleadership in the planning and conduction of the course at Erice, to Laura Vergani for her invaluable and critical cooperation prior, during and after the Institute and publication of this volume and to Maria Raffaele for the typing and editorial assistance of these proceedings.

Claudio Nicolini

CONTENTS

CELL STRUCTURE AND THE REGULATION OF GENES CONTROLLING PROLIFERATION AND DIFFERENTIATION: THE NUCLEAR MATRIX AND CYTOSKELETON

Gary S. Stein, Jane B. Lian, Thomas A. Owen, Janet L. Stein, Melissa Tassinari, Andre van Wijnen, Leesa M. Barone, Victoria Shalhoub, Michael Aronow, Gerry Zambetti[+], Steven I. Dworetzky*, Shirwin Pockwinse, and Joost Holthuis@

Department of Cell Biology, University of Massachusetts Medical Center, 55 Lake Avenue North, Worcester, MA 01655
Present address: [+]Department of Biology, Lewis Thomas Lab, Princeton University, Princeton, NJ 08544
Present address: *CNS Biology, Bristol-Myers Squibb, 5 Research Parkway, Wallingford, CT 06492
Present address: @The Central Lab for Blood Transfusion, Amsterdam, The Netherlands

In this chapter and in the one which follows we will present concepts and experimental approaches associated with the relationship of proliferation to differentiation with emphasis on the contribution of cell structure to the regulation of cell growth and tissue specific gene expression. While these relationships are of broad biological relevance, we will focus primarily on development of the osteoblast phenotype with the understanding that analogous principles apply to the regulation of phenotype expression in general.

Initially, we will review several of the fundamental elements of the sequence of events by which cell growth contributes to the onset and expression of differentiation whereby cells undergo a developmental maturation process, progressively acquiring phenotypic properties associated with specialized cells. Then, we will examine gene regulation related to expression of the osteoblast phenotype within the context of cellular architecture: first considering the potential contribution of the nuclear matrix to modifications in transcription during osteoblast growth and differentiation; then evaluating the contribution of the cytoskeleton to the stability and translatability of mRNAs.

Molecular Basis of Human Cancer
Edited by C. Nicolini, Plenum Press, New York, 1991

THE GROWTH-DIFFERENTIATION RELATIONSHIP

General Features

Although the concept of a relationship between proliferation and differentiation has been viewed as a necessary component of the developmental process for more than a century, the experimental approaches have been largely descriptive and the results primarily correlative. However, the recent development of culture systems which support the differentiation of specialized cells has permitted the combined use of biochemical, molecular and ultrastructural approaches to address the relationships of cell growth to the expression of cell and tissue-specific phenotypic properties (Figure 1). The application of in situ methodologies is particularly important in approaching questions related to differentiation since these in vitro systems support the development of a tissue-like organization analogous to that which occurs in vivo necessitating an understanding of molecular signaling mechanisms at the single cell level. Examples of such in vitro systems include: pluripotent promyelocytic leukemia cells which develop the monocytic, macrophage, or granulocytic phenotype[1,2]; adipocytes[3,4]; myoblasts[5,6]; keratinocytes[7,8]; and osteoblasts[9-14].

There are several features of the proliferation/ differentiation relationships exhibited by these in vitro systems which bear striking analogies to tissue development in

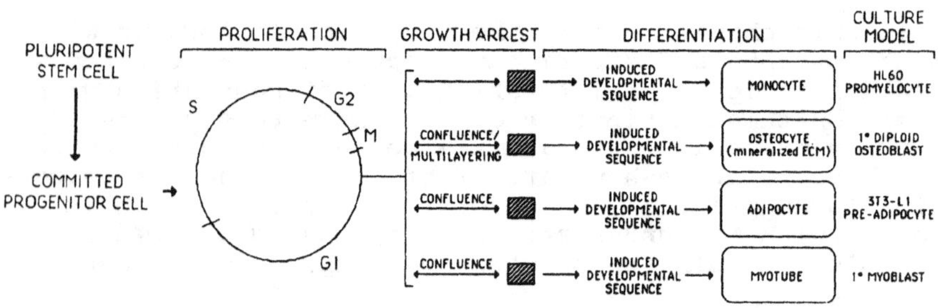

Fig. 1. Schematic representation of the events associated with progressive expression of monocyte, granulocyte, osteoblast, adipocyte and myotube phenotypes. Initially, the cells (HL-60 promyelocytic leukemia cells, primary cells of embryonic calvarial osteoblasts, 3T3-L1 pre-adipocytes or myoblasts) actively proliferate, expressing cell cycle and cell growth regulated genes, as well as genes encoding extracellular matrix proteins. Following growth arrest, a developmental sequence involving the sequential and selective expression of genes that results in the differentiated cell and tissue phenotype occurs. Completion of the proliferation period marks an important transition point where expression of tissue specific genes, often functionally related to the down-regulation of proliferation, is initiated.

vivo. In addition to validating the culture models, these relationships establish what appears to be important and possibly rate limiting steps in the differentiation process. Initially, a pluripotent stem cell and/or committed progenitor cell undergoes active proliferation, increasing the pool of precursor cells and the tissue mass to accommodate the biosynthesis of the specialized cell products required for intercellular and extracellular structural and functional properties unique to a developing tissue. While actively proliferating, these cells express genes encoding cell cycle and cell growth related proteins that support the complex and interdependent events associated with the proliferative process. Additionally the possibility should be considered, and results will be presented to support, that gene expression in proliferating cells may suppress expression of genes for later events in the differentiation process. Regulation of gene expression in proliferating cells that relate to cell cycle control is largely mediated at several post-transcriptional levels (e.g. mRNA stability and phosphorylation) providing the basis for a rapid response to accommodate cellular events which include DNA replication and mitotic division. In contrast, at the completion of proliferative activity the down-regulation of cell cycle related gene expression and the initiation of cell and tissue specific gene expression is largely controlled transcriptionally. Equally important, it should be emphasized that while the completion of proliferative activity in promyelocytes, osteoblasts, preadipocytes and myoblasts is an important point in the development of the tissue-specific phenotype, a progressive series of events occurs post-proliferatively that is necessary for the ordered development of the structural and functional characteristics of a differentiated cell -- each requiring a complex series of regulatory steps mediated at multiple levels.

THE OSTEOBLAST DEVELOPMENTAL SEQUENCE

A Temporal Pattern of Gene Expression

Normal diploid osteoblasts isolated from fetal rat calvaria (21-day rats) undergo an ordered developmental sequence during a 35 day culture period resulting in the formation of multilayered nodules of cells with a mineralized Type I collagen extracellular matrix[9-16] (Figure 2). By the combined use of molecular, biochemical, and ultrastructural analysis, the expression of cell growth and tissue-specific genes has been mapped during the progressive development of the bone cell phenotype within the context of the development of a bone tissue-like organization[12-14,16] (Figure 3). The temporal sequence of expression of genes encoding osteoblast phenotype markers in culture follows the pattern of gene expression and tissue distribution determined by *in situ* hybridization observed in neonatal long bones[17] and during fetal calvarial development *in vivo*[18] supporting the biological relevance of the osteoblast culture system (Figure 2).

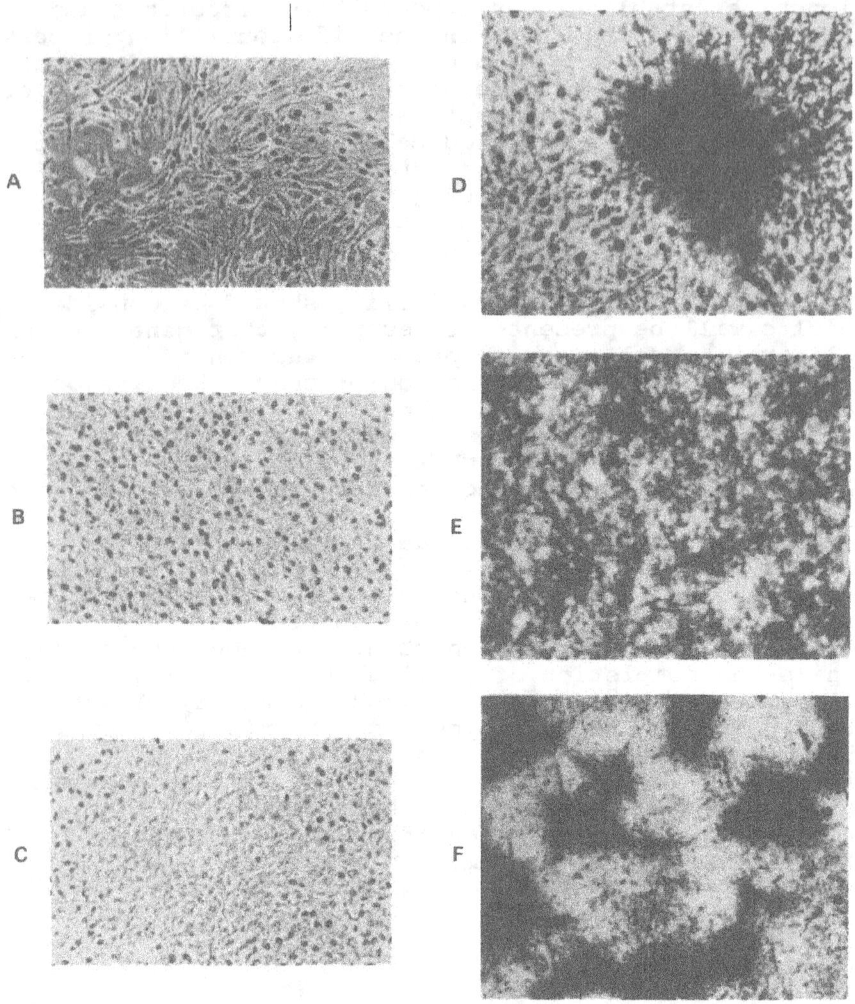

Fig. 2. Down-regulation of proliferation during the initial period of the development of the osteoblast phenotype in vitro reveals the initial cessation of proliferation in multilayered regions of the cultures. Note that on day 5 (Panel A) all cells are actively proliferating, on day 11 proliferative activity is still actively occurring throughout the cultures (Panel B), but that on day 12 there are multilayered regions of the culture where all cells have ceased proliferative activity with proliferation still ongoing in the non-multilayered regions (Panel C). The combination of alkaline phosphatase histochemistry and autoradiography following ^{3}H-thymidine incorporation shows that it is in multilayered regions of the culture that proliferation in first down-regulated and that it is these regions that first become intensely alkaline phosphatase positive (day 12, Panel D). The entire culture becomes alkaline phosphatase positive by day 16 (Panel E). The ordered deposition of mineral in nodules as extracellular matrix develops within the multilayered regions of the cells in the cultures is evident on day 35 (Panel F).

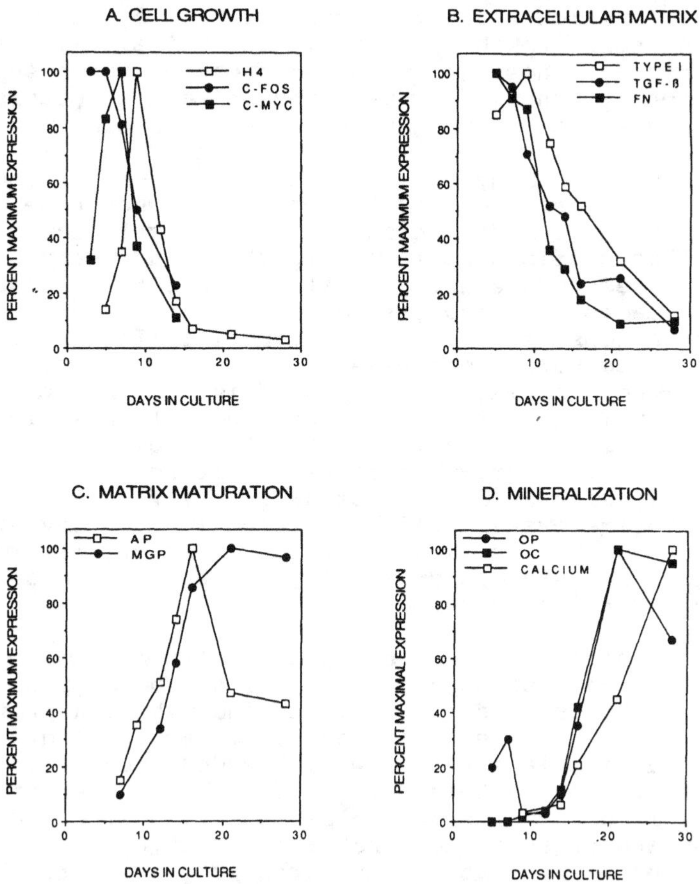

Fig. 3. *Temporal expression of cell growth, extracellular matrix, and osteoblast phenotype related genes during the development of the osteoblast phenotype <u>in vitro</u>. Isolated primary cells were cultured after confluence in BGJb medium supplemented with 10% FCS, 50 µg/ml ascorbic acid and 10mM β-glycerol phosphate. Cellular RNA was isolated at the times indicated (3,5,7,10,12,14,16,20,28, and 35 days) during the differentiation time course and assayed for the steady state levels of various transcripts by Northern blot analysis. The resulting blots were quantitated by scanning densitometry and the results plotted relative to the maximal expression of each transcript. (A) Cell growth related genes shown are H4 histone (reflects DNA synthesis), c-myc and c-fos. (B) Extracellular matrix associated genes represented are Type I collagen, fibronectin (FN) and transforming growth factor-β (TFG-β). (C) Genes associated with extracellular matrix maturation shown are alkaline phosphatase (AP) and matrix Gla protein (MGP). (D) Genes induced with extracellular matrix mineralization represented are osteopontin (OP) osteocalcin (OC) and calcium accumulation. Note the induction of alkaline phosphatase at the end of proliferation and the induction of osteocalcin and osteopontin with the onset of calcium deposition (day 12).*

Initially, during the first 10-12 days following isolation of osteoblasts from calvaria there is a period of active proliferation with expression of cell cycle (e.g. histone) and cell growth (e.g. c-myc and c-fos AP-1 activity) regulated genes. These genes encode proteins which support proliferation by functioning as transactivation factors in the case of c-myc and c-fos and as proteins which play the primary role in packaging newly replicated DNA into chromatin in the case of histones. During this proliferation period, and fundamental to development of the bone cell phenotype, several genes associated with formation of the extracellular matrix (Type I collagen, fibronectin and TGFβ) are actively expressed. These genes are gradually down-regulated during subsequent stages of osteoblast differentiation and the parallel relationship of TGFβ with Type I collagen gene expression in cultured osteoblasts[14] as well as during endochondral bone formation <u>in vivo</u>[19,20] is consistent with a major role for TGFβ in regulating extracellular matrix biosynthesis. It is primarily during the proliferation period that the activity of growth factors, their regulators and the associated signal transduction mechanisms influence the osteoblast parameters and this has recently been reviewed by Centrella, et al.[21]. Modifications of proliferation related genes, e.g. c-fos expression <u>in vivo</u> (in transgenic animals) results in altered bone formation[22].

With the decline in DNA synthesis (^{3}H-thymidine incorporation and histone gene expression) the expression of alkaline phosphatase (enzyme activity and mRNA), a protein associated with the bone cell phenotype, increases greater than ten-fold immediately following the down-regulation of proliferation. During this period (from 12-18 days) the extracellular matrix undergoes a series of modifications in composition and organization that renders it competent for mineralization. Then, as the cultures progress into the mineralization stage, cellular levels of alkaline phosphatase mRNA decline. Two other bone related genes, osteopontin[23-25] and osteocalcin[26,27], exhibit a different pattern of expression. Osteocalcin is not detectable prior to day 12 in culture and does not reach a significant level of expression until 16-20 days after isolation when expression increases coordinately with total mineral accumulation. Osteopontin similarly reaches peak levels of expression during the mineralization period (days 16-20). It is also expressed during the period of active proliferation at 25% of maximal levels. This is not an unexpected result for two reasons. First, osteopontin expression during the proliferative period is consistent with the increased level of expression during the pre-replicative phase of the cell cycle following serum stimulation of quiescent fibroblasts and following oncogene transformation or of phorbol ester treatment of fibroblasts[28]. Here one can speculate that the proliferation and tumorgenic-related function of osteopontin may be related to control of relationships between cells and extracellular matrices[28]. Second, the induced expression of osteopontin coincident with mineralization may be related to physical properties of the protein. Osteopontin is a 60 kd acidic glycoprotein[24] containing o-phosphoserine, thereby possessing several putative calcium binding sites which are known to be important for cell proliferation and also for mineralization of the extracellular matrix in bone[29]. It therefore appears that expression of the osteopontin gene early and late in the osteoblast developmental sequence may be mediated by alternative regulatory mechanisms. Not to be overlooked is the possibility that the lower levels of osteopontin mRNA observed during the proliferation period may in part reflect mRNA transcribed in vivo in osteoblasts undergoing matrix mineralization prior to the isolation from fetal calvaria. The general structural and biological properties of non-collagenous proteins which are associated with the bone cell extracellular matrix has recently been reviewed by Heinegard[25].

Yet another category of genes expressed during the osteoblast developmental sequence is represented by matrix Gla protein. The matrix Gla protein gene is vitamin D responsive and the encoded 10 kd polypeptide which is associated with the osteoblast extracellular matrix contains five γ-carboxyglutamic acid residues added posttranslationally by a vitamin K dependent microsomal carboxylase. In contrast to many of the other genes expressed during specific periods of the osteoblast developmental sequence, matrix Gla protein is initially expressed in proliferating osteoblasts and expression continues during extracellular matrix maturation and mineralization. Interestingly, matrix Gla protein is not found exclusively in osteoblasts, but is also abundant in chrondrocytes and in several nonskeletal tissues with extensive extracellular matrices (e.g. lung and kidney).

The observed temporal expression of osteoblast phenotype properties during the developmental sequence is a reflection of functional activities necessary for the progressive formation of bone tissue. The expression of alkaline phosphatase mRNA and enzyme activity prior to the initiation of osteoblast mineralization suggests that alkaline phosphatase may be involved in preparation of the extracellular matrix for the ordered deposition of mineral and that the co-expression of other genes such as osteocalcin and osteopontin may support the onset and progression of extracellular matrix mineralization. Alternatively, the induction of these mineralization-associated genes may reflect an acquisition of osteoblast properties necessary for signaling bone turnover in vivo[30]. Taken together, the patterns of expression of these genes and the synthesis of the encoded proteins, determined biochemically and by histochemical staining, demonstrate that a temporal sequence of gene expression exists during the culture period associated with development of the extracellular matrix and reflects maturation of the osteoblast phenotype in vitro.

The biological relevance of the temporal expression of osteoblast parameters in vitro to bone formation in vivo is demonstrated by the fidelity of the tissue-like organization developed at the completion of the in vitro osteoblast developmental sequence. This is reflected by intense von Kossa silver staining of the mineralized nodules indicating hydroxyapatite deposition (Figure 2F). The bone-tissue-like organization in these cultures is further supported by comparison of the ultrastructure of the mineralized regions of the culture. Sections through an intact 21 day fetal rat calvarium exhibits a similar ordered deposition of crystals within and between the orthogonally organized bundles of collagen fibrils. No evidence for cell necrosis or intracellular calcification is indicated in the cultures, particularly where mineralized matrix has enveloped the osteoblasts. In heavily mineralized mature cultures (after day 35), a pattern of gene expression is found analogous to the mature osteocyte in osseous tissue. In vitro alkaline phosphatase and collagen mRNA levels are almost nondetectable and osteopontin and osteocalcin levels have declined; in vivo alkaline phosphatase histochemistry indicates activity less than that found at the mineralizing front and active collagen biosynthesis is not associated with osteocytes.

Evidence for a Functional Relationship of Cell Growth to Expression of the Osteoblast Phenotype

By combining ^{3}H-thymidine labeling and in situ autoradiography with alkaline phosphatase histochemistry, it has been possible to directly establish the relationship between proliferation and initiation of tissue-specific gene expression at the single cell level during the osteoblast developmental sequence[14]. As seen in Figure 2, it is apparent that proliferation initially ceases in the discrete multilayered foci which form throughout the osteoblast cultures (Figure 2C) and it is these cells in the multilayered nodules which first express alkaline phosphatase (Figure 2D). This is in contrast to proliferating cells in the internodular regions of the cultures where alkaline phosphatase activity is not observed until several days later when the entire cultures consist of multilayered non-proliferating cells (Figure 2E).

These results confirm on an individual cell basis that a temporal sequence of gene expression occurs and that at least some events (proliferation and alkaline phosphatase expression) appear to be sequential, mutually exclusive events in the same cell, i.e., proliferation must be down-regulated prior to the expression of alkaline phosphatase. This transition from a proliferating cell to one which can express an early marker of the osteoblast phenotype (alkaline phosphatase) represents the first restriction point where cessation of proliferation appears to be required for initiation of tissue-specific gene expression associated with the distinctive characteristic features of bone -- formation of the mineralized extracellular matrix.

Another direct demonstration that the down-regulation of proliferation induces the expression of some genes which are normally expressed later in the osteoblast developmental sequence is derived from experiments which establish that inhibition of DNA synthesis in actively proliferating osteoblasts results in a rapid and selective down-regulation of cell growth genes[14] (Figure 4). This is paralleled by a four fold increase in alkaline phosphatase mRNA levels, indicating that the premature down-regulation of proliferation induces the expression of an early marker for the extracellular matrix maturation period of the osteoblast

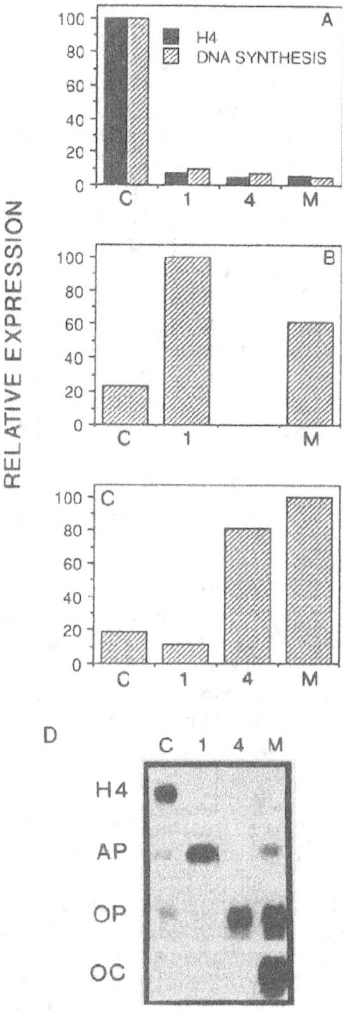

Fig. 4. *Coupling of alkaline phosphatase and osteopontin but not osteocalcin expression to the down-regulation of proliferation. Proliferation was inhibited in actively growing osteoblasts (day 5) by addition of 5 mM hydroxyurea (HU). Following HU addition, cells were harvested at 1 and 4 hr., examined for DNA synthesis (A), and cellular RNA prepared and analyzed for H4 histone (A), alkaline phosphatase (B) and osteopontin (C). The Northern blots of these transcripts and osteocalcin is also shown (D). Note that osteocalcin, which is not present in 5 day cultures (C), is also not induced following inhibition of proliferation by HU. For comparison, the relative expression of these genes in non-HU-treated control (C) and in mineralized (M) cultures (30 days after plating) is also shown.*

developmental sequence. Increased expression of alkaline phosphatase with decreased proliferative activity has similarly been observed in ROS 17/2.8 cells[31]. With inhibition of DNA synthesis, levels of osteopontin mRNA also increase to levels which approximate those present during the extracellular matrix mineralization period of cultured osteoblasts. These data suggest a direct functional coupling of the down-regulation of proliferation at the first transition point early during the osteoblast developmental sequence with the preferential expression of genes which are normally induced in cells with a mature extracellular matrix. However, osteocalcin which is expressed in osteoblasts late during the period of extracellular matrix mineralization, is not induced by simply inhibiting proliferative activity[14]. The absence of osteocalcin induction is consistent with the concept that there is at least a second set of genes whose expression is not directly coupled to the down-regulation of proliferation, but rather to development of the more differentiated osteoblasts in a mineralized matrix. These experiments in which premature differentiation has been promoted by inhibition of proliferation reveal a second transition point in the developmental sequence of osteoblast differentiation, since inhibition of proliferation supports expression of genes that are expressed only during progression of the developmental sequence up to the stage where mineralization is initiated. Mineral deposition may be required to signal expression of a subset of osteoblast phenotype genes, such as osteocalcin. Additionally, other genes expressed during the period of extracellular matrix maturation which may be required to render the matrix competent for mineralization may also not be induced by inhibition of proliferation.

The biological significance of the two transition points in the osteoblast developmental sequence is further suggested by the relationship between mineralization and the sequential expression of genes during the progressive development of the osteoblast phenotype. When cells are maintained under conditions which support extracellular matrix mineralization, osteocalcin mRNA and biosynthesis increase steadily beginning at day 15 in parallel with calcium accumulation in the cell layer. In contrast, when cultures are maintained under non-mineralizing conditions, calcium does not begin to accumulate in the cell layer and osteocalcin gene expression does not occur until approximately day 25; however, the presence or absence of mineralization has no effect on the expression of genes occurring during the proliferative period, on passage through the first transition point or the onset of alkaline phosphatase expression. These experiments provide additional evidence to support the existence of the second transition point, since the cells can progress through the proliferation and extracellular matrix maturation sequence to the onset of mineralization, but cannot initiate expression of genes related to the mineralization stage unless mineral accumulation occurs. Genes such as osteocalcin are not only temporally expressed late in the osteoblast developmental sequence, but are also "coupled" to deposition of hydroxyapatite.

A Model for the Relationship of Cell Growth to the Onset and Progression of Osteoblast Differentiation

The results we have presented based on determinations of molecular, biochemical and histochemical parameters are consistent with a reciprocal and functional relationship between proliferation and a sequential development of the osteoblast phenotype which is schematically illustrated in Figure 5. The progressive and interdependent series of biochemical events that characterizes the osteoblast developmental sequence reflects the selective expression, initially of genes encoding cell growth and extracellular matrix proteins and subsequently a series of tissue-specific genes[14]. Such modifications in the temporal pattern of gene expression, reflected by both the activation and repression as well as by the extent to which specific genes are expressed, are the basis for proposing a developmental sequence with three distinct periods: proliferation, extracellular matrix maturation and mineralization. This pattern of gene expression suggests two principal transition points in the osteoblast developmental sequence where important regulatory signals may be required for the progressive expression of the bone cell phenotype to proceed: the first when proliferation is down-regulated and gene expression associated with extracellular matrix maturation is initiated, and the second transition at the onset of mineralization.

Although unquestionably a simplification of an extremely complex series of biological interactions, the temporal pattern of expression suggests a working model for the relationship between growth and differentiation whereby genes involved in the production and deposition of the extracellular matrix must be expressed during the proliferative period for the onset and progression of differentiation to occur. One can postulate that proliferation is functionally related to the synthesis of a bone specific extracellular matrix and that the maturation and organization of the extracellular matrix contributes to the shutdown of proliferation which then promotes expression of genes that render the matrix competent for mineralization, a final process that is essential for complete expression of the mature osteoblast phenotype. The onset of extracellular matrix mineralization and/or events early during the mineralization period may be responsible for the down-regulation of genes expressed during extracellular matrix maturation and organization. Clearly, in this model the development of an extracellular matrix is integrally related to the differentiation stages and numerous studies have shown enhancement of osteoblast phenotype properties in the presence of ascorbic acid (summarized in references[13,14]) and under a variety of conditions which promote extracellular matrix biosynthesis and organization (see the next chapter). This working model provides a basis for addressing whether particular stages of osteoblast differentiation exhibits selective responsiveness to actions of hormones and other physiologic factors that influence osteoblast activity and other questions related to the molecular mechanisms associated with bone formation.

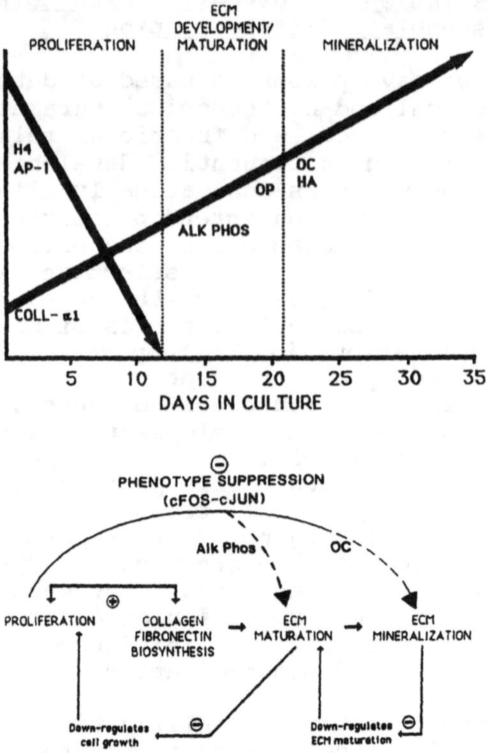

Fig. 5. *A model of the relationship between proliferation and differentiation during the rat osteoblast developmental sequence. This relationship is schematically illustrated within the context of modifications in expression of cell cycle and cell growth regulated genes, as well as genes associated with the maturation, development and mineralization of the osteoblast extracellular matrix. The three principal periods of the osteoblast developmental sequence are designated within broken vertical lines (proliferation, matrix development and maturation, and mineralization). Commitment periods and restriction points which are indicated were experimentally established. A functional relationship between the down-regulation of proliferation and the initiation of extracellular matrix development and maturation is based on stimulation of alkaline phosphatase and osteopontin gene expression when proliferation is inhibited but the developmental sequence is induced only to the second transition point. Growth of the osteoblast under conditions which do not support mineralization confirms the day 20 restriction point since the developmental sequence proceeds through the proliferation and the extracellular matrix development/maturation periods, but not further.*
The lower panel schematically illustrates our model for suppression of marker genes of the developing alkaline phosphatase and the mature osteoblast phenotype (osteocalcin) in actively proliferating cells by protein binding to AP-1 Sites which reside within the vitamin D responsive element of both genes and additionally to the CCAAT element of the osteocalcin gene promoter. The proliferation period

Molecular Mechanisms Operative at the Proliferation/ Differentiation Transition Point

Molecular mechanisms operative at the transition point early in the osteoblast developmental sequence, when proliferation ceases and genes associated with extracellular matrix maturation and specialization are induced, can be addressed by examining the down-regulation mechanisms of genes expressed during the proliferation period. Examples of such genes are those encoding histone proteins which package newly replicated DNA into chromatin. Histone gene expression is restricted to the S phase of the cell cycle and is tightly coupled with DNA replication[32,33]. At the completion of the proliferation period, histone gene expression is down-regulated at the transcriptional level and a systematic examination of protein-DNA interactions in the proximal regulatory elements of cell cycle regulated H4 and H3 histone genes indicates that this down-regulation of transcription is mediated by a selective loss in the binding of a transcription factor HiNF-D to sequences which influence both specificity and level of transcription (Site II) (Figure 6a and b)[34-36]. A functional relationship between loss of HiNF-D - Site II interactions and the onset of osteoblast differentiation is further suggested by the persistence of Site II occupancy by HiNF-D when cell proliferation is inhibited under conditions which do not promote the activation of genes associated with extracellular matrix maturation and specialization, i.e., conditions where cell growth is blocked but the progressive differentiation of the osteoblast does not proceed[34].

Another molecular marker for the transition point between completion of proliferation and the onset of gene expression characteristic of the osteoblast phenotype (e.g. alkaline phosphatase) is expression of cell cycle independent histone genes which encode high molecular weight poly A+ mRNAs[37]. Expression of poly A+ histone mRNAs at the onset of tissue specific gene expression has similarly been observed during monocytic differentiation in HL-60 promyelocytic leukemia cells[38] as well as during adipocyte differentiation. The encoded histone proteins may be involved in the remodeling of

---

supports the synthesis of a Type I collagen/ fibronectin extracellular matrix which continues to mature and mineralize. The formation of this matrix down-regulates proliferation and matrix mineralization down-regulates the expression of genes associated with the extracellular matrix formation-maturation period. Occupancy of AP-1 sites by the oncogene-encoded Fos and Jun proteins, which are actively synthesized in proliferating osteoblasts, suppresses alkaline phosphatase expression during proliferation. The coordinate occupancy of the AP-1 Sites in the osteocalcin box (CCAAT-containing proximal promoter element) and the vitamin D responsive element of the osteocalcin gene by Fos-Jun and/or related proteins suppresses the basal and vitamin D enhanced expression of the osteocalcin gene prior to the initiation of osteocalcin basal expression at the onset of extracellular matrix mineralization. AP-1, AP-1 binding activity; H4, H4 histone; COLL-αI, Type αI collagen; ALK PHOS, alkaline phosphatase; OP, osteopontin; OC, osteocalcin; HA, total accumlated hydroxyapatite (calcium + phosphate).

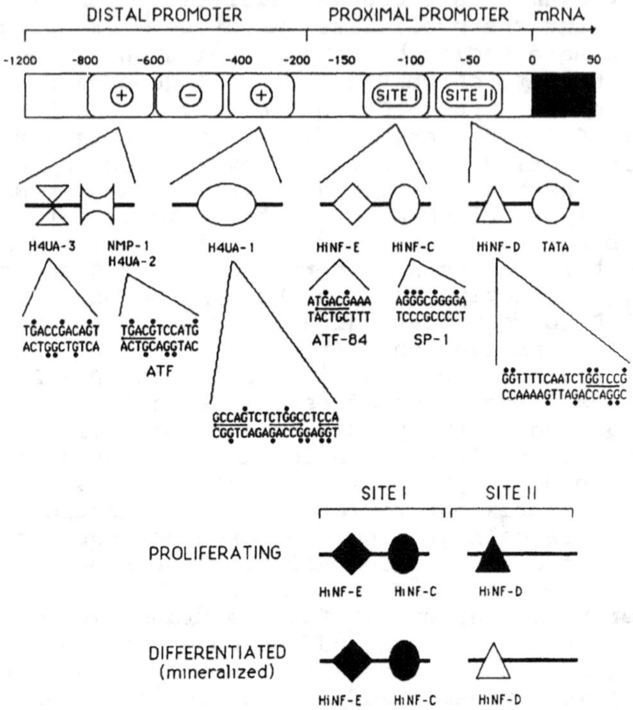

Fig. 6A. *Regulatory sequences and molecules involved in the regulation of histone gene expression during proliferation of osteoblasts and following down-regulation of expression with the onset of differentiation. Schematically illustrated is the modular organization of a cell cycle/cell growth regulated H4 histone gene that is expressed during osteoblast proliferation. Two segments of the promoter designated Site I and Site II which influence specificity and levels of transcription are indicated in the diagram of the gene. Promoter binding factors HiNF-E and HiNF-C bind to Site I and HiNF-D binds to Site II. In proliferating cells, occupancy of the two sites by the three factors is observed (solid symbols). In contrast, the shut-down of proliferation that occurs when osteoblasts initiate expression of genes related to extracellular matrix maturation and specialization (e.g. alkaline phosphatase) is associated with a selective loss of HiNF-D - Site II interactions (open symbols) while occupancy of Site I by HiNF-E and HiNF-C persists.*

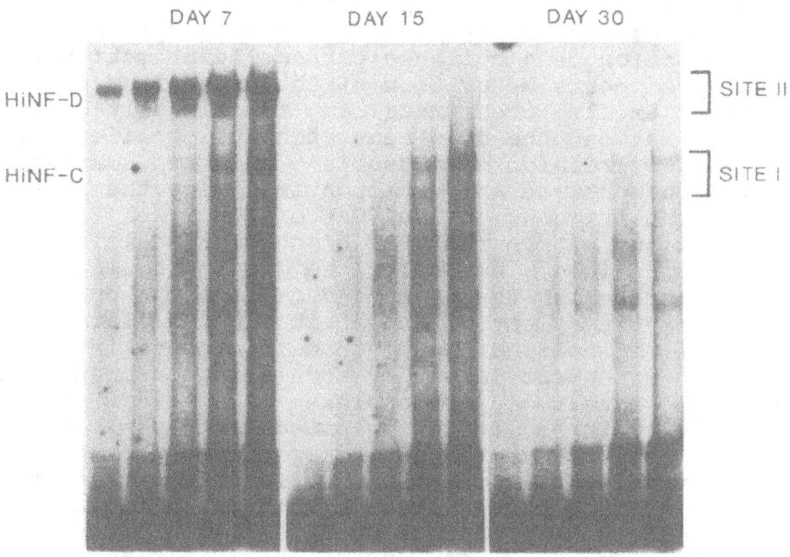

Fig. 6B. *Selective loss of factor HiNF-D only when proliferation is down-regulated as a part of a developmental sequence. (A) Primary fetal rat osteoblasts were harvested on either day 7 (proliferating), 15 (non-proliferating, undergoing extracellular matrix maturation) or day 30 (mineralized) after plating for nuclear protein extracts which were analyzed for HiNF-D DNA binding activity. (B) Nuclear proteins were extracted from either confluent serum starved osteoblasts or from control osteoblast cultures and analyzed for HiNF-D DNA binding activity.*

chromatin architecture which occurs following the completion of proliferation and may be necessary to support expression of genes associated with the development of tissue specific phenotypic properties and the specialized functions of the differentiated cells.

DEREGULATION OF THE RELATIONSHIP BETWEEN CELL GROWTH AND TISSUE SPECIFIC GENE EXPRESSION IN OSTEOSARCOMA CELLS

The developmental sequence associated with the differentiation of normal diploid osteoblasts, both *in vitro* and *in vivo*, supports two concepts related to the progressive development of the osteoblast phenotype. First, the proliferative period supports collagen gene expression and initial production of a collagen extracellular matrix that subsequently supports maturation of the osteoblast phenotype. Second, there exists a reciprocal and functionally coupled relationship between the down-regulation of proliferation and initiation of expression of osteoblast phenotype markers such as alkaline phosphatase and osteopontin. Here the presence of an AP-1 site, contiguous to the VDRE of the Type I collagen gene promoter and within the VDRE of the osteocalcin gene promoter [39,40] provides a basis for the growth regulated AP-1 site binding proteins (the c-jun and c-fos proto-oncogene encoded polypeptides) to influence vitamin D-mediated expression of the collagen and osteocalcin genes in a proliferation-dependent manner by modulating occupancy of the VDRE by the hormone-receptor complex. High levels of AP-1 binding activity in proliferating osteoblasts can block occupancy of the VDRE in the osteocalcin gene promoter, thereby suppressing transcription; while in contrast, the absence of overlap by the AP-1 site with the vitamin D receptor binding domain in the Type I collagen promoter VDRE is compatible with expression during proliferation[41] (See Figure 5).

While there are density dependent effects on gene expression in transformed osteoblasts and osteosarcoma cell lines, the three defined periods which characterize the developmental sequence observed in cultured normal diploid osteoblasts, with a specific pattern of gene expression and evidence for transition points where signaling mechanisms necessary for the progressive expression of the osteoblast phenotype occur, are not operative. In several transformed osteoblast and osteosarcoma cell lines which have been extensively examined, for example in the ROS 17/2.8 rat osteosarcoma cell line, there is a deregulation of the sequential pattern of gene expression. This relaxation of control mechanisms which permits genes that are sequentially expressed in diploid osteoblasts[14] to be expressed simultaneously is reflected by the concomitant expression of tissue specific genes such as alkaline phosphatase, osteopontin and osteocalcin while the cells are actively proliferating[42,43] (Figure 7). Equally important, perturbations in the signaling mechanisms which interface the down-regulation of cell growth and induction of genes which support extracellular matrix maturation and specialization are implicated. Consistent with such reasoning, we have recently

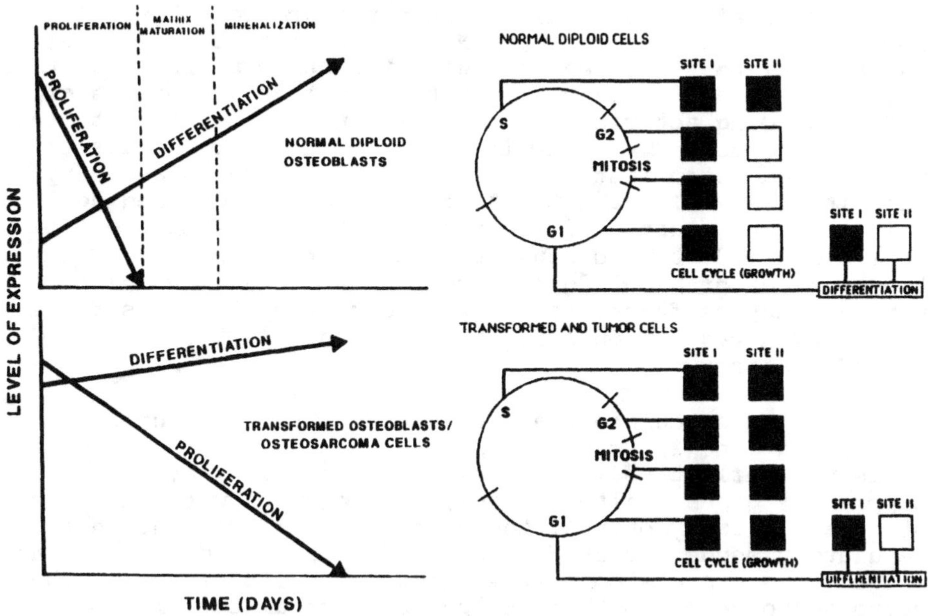

Fig. 7. *Schematic representation of the relationship between growth and differentiation during the osteoblast developmental sequence in normal diploid osteoblasts and in transformed osteoblasts or osteosarcoma cells. The broken vertical lines in the upper diagram indicate the two principal transition points in the developmental sequence exhibited by normal diploid osteoblasts during the progressive acquisition of the bone cell phenotype; the first at the completion of proliferation when genes associated with matrix development and maturation are upregulated and the second at the onset of extracellular matrix mineralization. Note that the constitutive expression of osteoblast phenotype markers in the transformed osteoblasts and osteosarcoma cells reflects the absence of these two developmentally important transition points and the apparent loss of signaling mechanisms that are operative at these stages of the osteoblast developmental sequence in normal diploid cells.*

The proliferation vectors reflect the level of cell growth and expression of cell cycle and proliferation related genes. The differentiation vectors reflect the expression of genes associated with the biosynthesis, maturation and mineralization of the extracellular matrix. In the normal diploid osteoblasts there are three distinct periods to the developmental sequence with extracellular matrix biosynthesis an early event occurring primarily during the proliferation period and the maturation, specialization and mineralization of the extracellular matrix occurring after the completion of proliferation. In contrast, the relationship between cell growth and expression of genes encoding osteoblast phenotype markers is not apparent in transformed osteoblast or osteosarcoma cells; cell growth and tissue specific gene expression occurs concomitantly and the relationship between growth and differentiation is deregulated.

(continued)

observed deregulation of the cell cycle dependent histone gene promoter binding factor HiNF-D in osteosarcoma cells. In normal diploid osteoblasts, HiNF-D binding activity is cell cycle regulated. Nuclear protein extracts prepared from these cells in S phase contain distinct and measurable levels of HiNF-D binding activity, while this activity is not detectable in G1 phase cells[44]. In contrast, in osteosarcoma cells or transformed osteoblast cell lines, HiNF-D binding activity is constitutively elevated throughout the cell cycle and declines only with the onset of differentiation[34,44]. The change from cell cycle mediated to constitutive interaction of HiNF-D with the promoter of a cell growth controlled gene is consistent with and may be functionally related to the loss of stringent cell growth regulation associated with neoplastic transformation.

One can speculate that the deregulation of a growth controlled gene in transformed osteoblasts and in osteosarcoma cells may reflect modifications in the activity of tumor suppressor genes. Alternatively, or together with the loss of stringent growth control, there may be modifications in the regulatory sequences and/or in the factors which control the progressive expression of the tissue-specific genes and their response to cell growth or morphogenic regulatory factors. While the specific molecular mechanisms remain to be established, by further understanding the deregulation of growth in bone tumors we can anticipate gaining additional insight into control of the tightly coupled relationship between proliferation and development of the osteoblast phenotype.

THE NUCLEAR MATRIX AND TRANSCRIPTIONAL CONTROL

A key question that must be addressed to understand regulation of gene expression at the transcriptional level is how with a low representation of promoter binding factors and sequence specific regulatory elements in the nucleus, a threshold concentration can be achieved to initiate RNA synthesis. Here, the answers may at least in part reside in the three dimensional organization of the cell nucleus, where the nuclear matrix exhibits properties which are consistent with a functional involvement in the control of transcription.

Fig. 7. (cont.) *The right panels indicate schematically the occupancy of Site I and Site II by sequence specific promoter binding factors. Protein-DNA interactions are indicated by filled boxes. Note that protein-DNA interactions occur at Site I throughout the cell cycle in proliferating cells (normal diploid and transformed/tumor) as well as following differentiation. However, at Site II, occupancy by promoter binding factors is S-phase specific in normal diploid cells and constitutive throughout the cell cycle in transformed/tumor cells. Only following differentiation are protein-DNA interactions at Site II lost in transformed and tumor cells. These findings indicate a stringent cell cycle regulation of Site II protein-DNA interactions in normal diploid cells which are deregulated (constitutive) following transformation.*

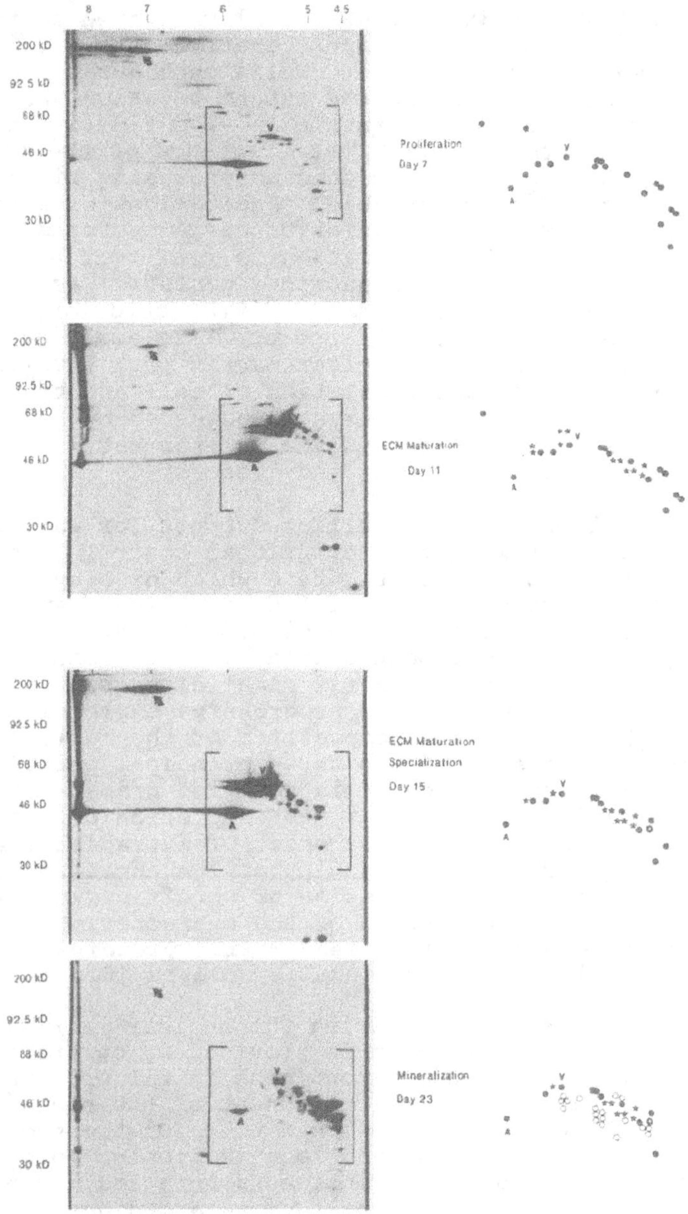

Fig. 8. *Progressive changes in the protein composition of the nuclear matrix during osteoblast differentiation. The fluorographs represent 2-dimensional gel electrophoresis of ^{35}S-methionine labeled nuclear matrix proteins isolated from days 7, 11, 15, and 23 primary rat osteoblast cultures. The schematic focus on the bracketed region shows stage-specific changes that occur in the nuclear matrix composition. The arrow points to a large 190kd protein that completely disappears as the cells differentiate. Symbols represent those proteins that appears on that particular day. •-day 7; ★-day 11, ⊗-day 15, ○-day 23. The symbols are removed when the given protein is no longer present.*

The nuclear matrix is operationally defined as the proteinaceous nuclear substructure that resists both nuclease digestion and salt extractions. While considerable controversy exists regarding the extent to which various nuclear matrix preparations retain _in vivo_ fidelity of composition and organization[45] the existence of the nuclear matrix as a network of polymorphic anastomosing filaments within the nucleus is undeniable. Indications of a role for the nuclear matrix in the regulation of gene expression includes: sites for DNA replication [46]; preferential association with actively transcribed genes[47,48]; association with HN-RNA[49]; RNA synthesis at fixed transcriptional complexes; pre-mRNA splicing[50]; and specific association of steroid receptors with the nuclear matrix[51,52]. Involvement of the nuclear matrix in the regulation of cell and tissue specific gene expression is further suggested by recent demonstrations of variations in the nuclear matrix protein composition of different cells and tissues[53-57].

Additional and more compelling support for participation of the nuclear matrix in transcriptional control is provided by the results presented in Figure 8 which displays the two dimensional electrophoretic profiles of nuclear matrix proteins analyzed throughout the osteoblast developmental sequence. It is evident that changes in the protein composition of the nuclear matrix parallel sequential expression of genes during the progressive expression of the osteoblast phenotype. The composition of the nuclear matrix is constant within each of the three principal periods of osteoblast differentiation but is modified dramatically at the two key transition points -- at the completion of the proliferation period and at the onset of extracellular matrix mineralization. This relationship between nuclear matrix protein composition and expression of specific genes is further supported by retention of the characteristic stage-specific representation of nuclear matrix proteins when the osteoblast developmental sequence is delayed (Figure 9).

More direct evidence linking the nuclear matrix with the regulation of gene expression is provided by our recent demonstration that the H4 histone gene distal promoter binding factor NMP-1 (Figure 10) is a component of the nuclear matrix. These results are consistent with a role for the nuclear matrix in the concentration and localization of both actively transcribed genes and transcription factors and may explain how, with low representation of factors and sequence specific regulatory elements in the nucleus, a threshold to initiate transcription may be obtained. In a broader biological context, these results provide a basis for understanding how transcriptional control can be mediated by the nuclear architecture.

THE CYTOSKELETON AND POST-TRANSCRIPTIONAL REGULATION

Modulation of mRNA stability plays an important role in the regulation of gene expression during growth and differentiation providing a mechanism to "fine tune" cellular levels of mRNA which can then serve as rate limiting steps in

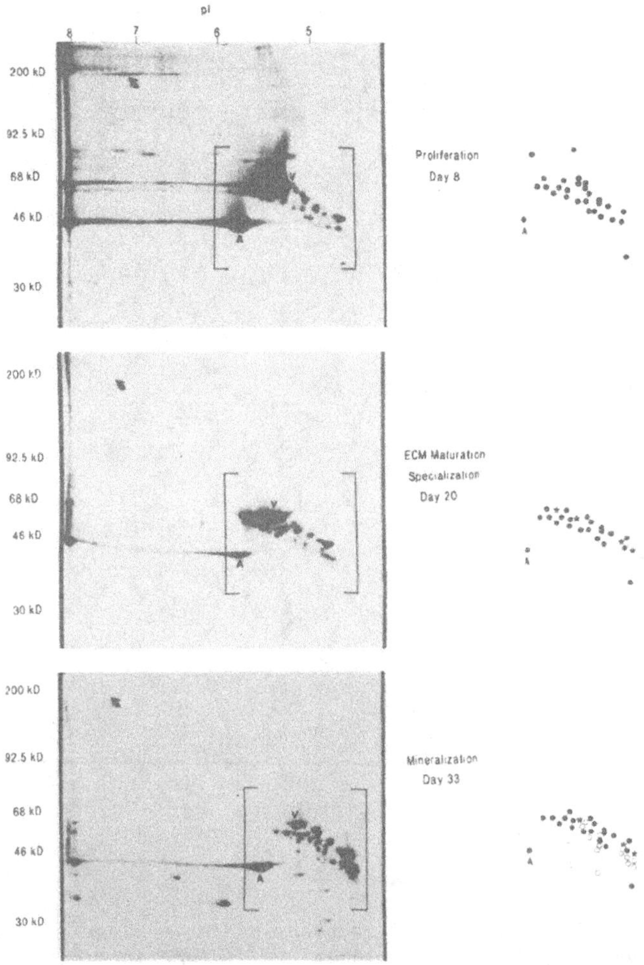

Fig. 9. *Nuclear matrix composition of rat osteoblasts during delayed differentiation. Cells were cultured in the absence of β-glycophosphate which delays the onset and extent of mineralization. The fluorographs represent 2-dimensional gel electrophoresis of ^{35}S-methionine labeled nuclear matrix proteins from days 8, 20, and 33. Symbols are those proteins that appeared on that particular day. •-day 8, ★-day 20, o-day 33. Note the pattern on day 33 resembles that seen on day 23 during the normal differentiation course (Fig. 8). These results support the effectiveness of utilizing nuclear matrix protein profiles as a marker of specific stages of the osteoblast developmental sequence.*

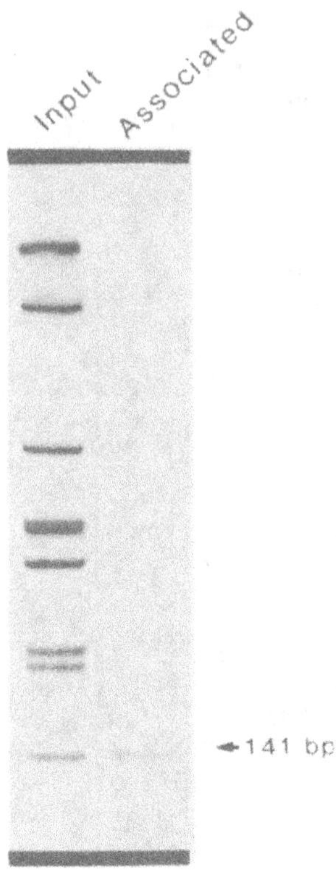

Fig. 10. *Sequence-specific interactions of the nuclear matrix protein NMP-1 with a regulatory element in a human H4 histone gene promoter.*

(A) Localization of an <u>in vitro</u> matrix-associated region within the H4 histone gene promoter. A mixture of radiolabeled cloned histone gene promoter segments and pUC19 fragments are incubated in the presence of unlabeled E. coli competitor DNA. A specific 141 bp DNA fragment is seen to be preferentially associated with the nuclear matrix. This fragment is located between -589 and -730 upstream from the cap site.

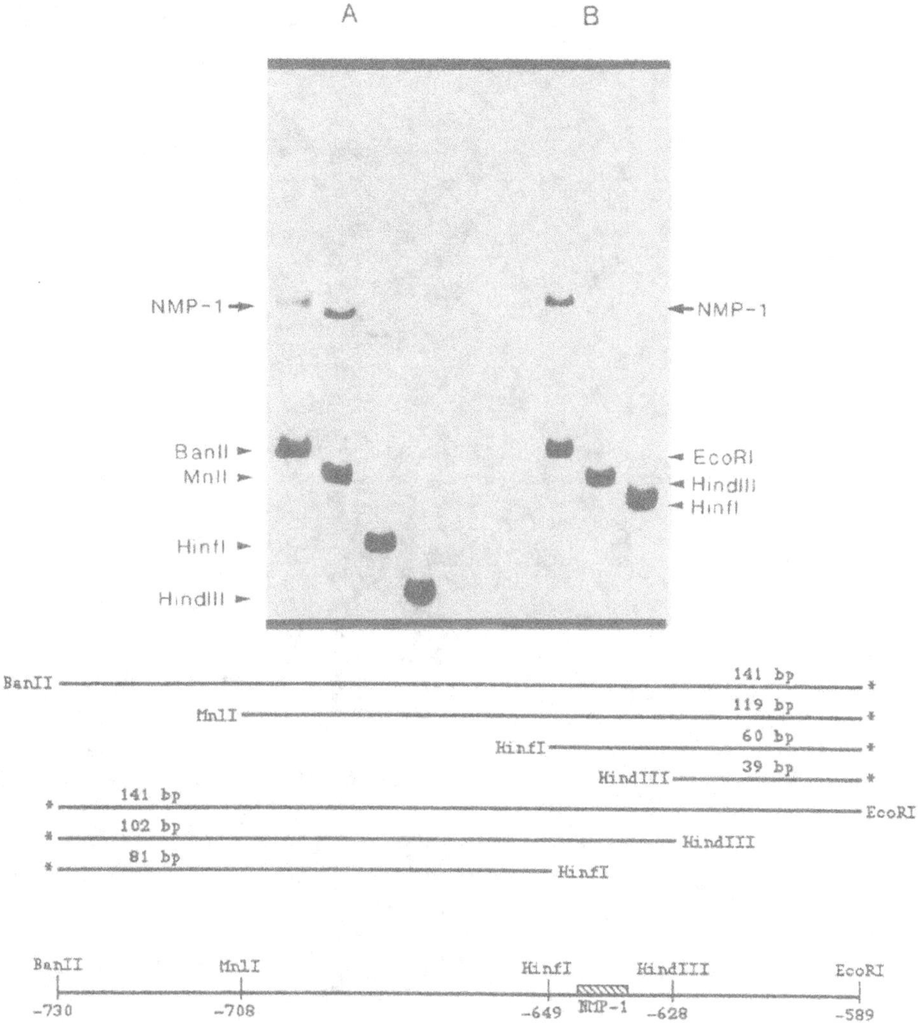

Fig. 10. *(B) Nuclear matrix protein-DNA interaction within the 141 bp distal promoter element. A gel mobility shift assay in conjunction with a bidirectional deletion analysis was carried out to determine the region within the 141 bp fragment where the site of protein-DNA interaction occurs. It is demonstrated that nuclear matrix proteins(s), NMP-1, bind to this distal element between the HiNF-1 and HiND-III restriction sites.*

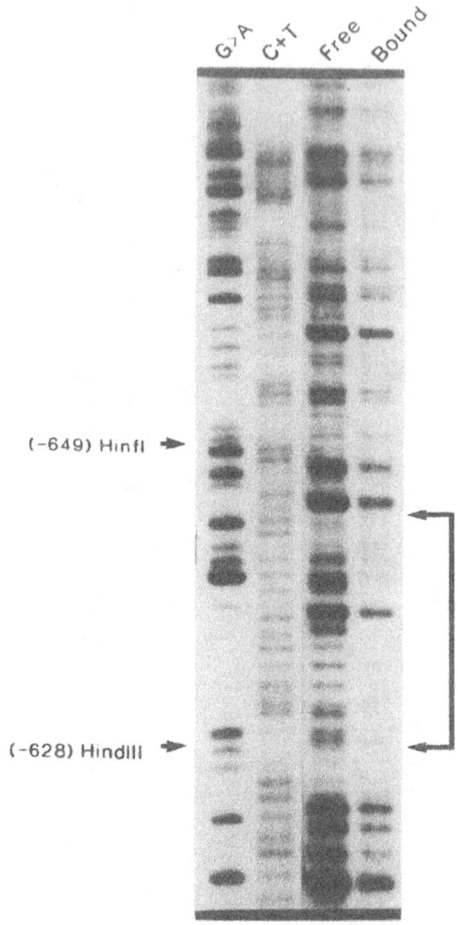

Fig. 10. *(C) Op-Cu footprint analysis of NMP-1 binding. The sequence specific interaction of NMP-1 was determined by footprint analysis. Comparison of the free and bound lanes shows that NMP-1 protects the DNA in between the HiNF-1 and HiND-III restriction sites.*

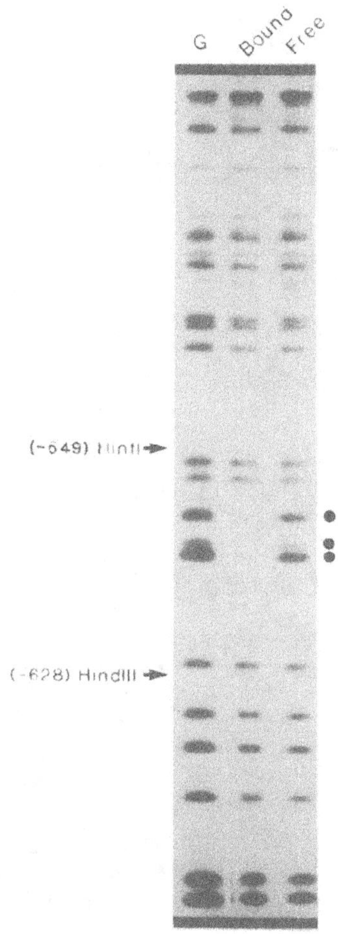

Fig. 10. *(D) Identification of specific guanine residue contacts with NMP-1. Specific nucleotide contacts of NMP-1 were established by DMS protection of the gel shifted protein-DNA complex. The recognition binding sequence is GGA<u>CGTCA</u>. The underlined segment contains the core recognition binding sequence for the ATF family of transcription factors. F, Free probe; B, Bound prove, G, G ladder.*

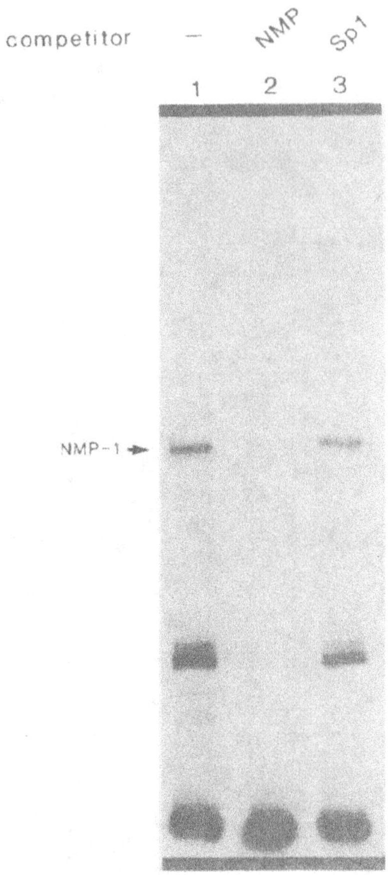

Fig. 10. (E) Competition analysis of the NMP-1/DNA complex. Oligonucleotides synthesized for the NMP-1 and Sp1 binding sequences were used as competitors in the gel mobility shift assay. The oligonucleotides were added in 250 molar excess. The NMP oligonucleotide specifically competes out the NMP-1 interaction, whereas the Sp1 oligonucleotide does not have any effect.

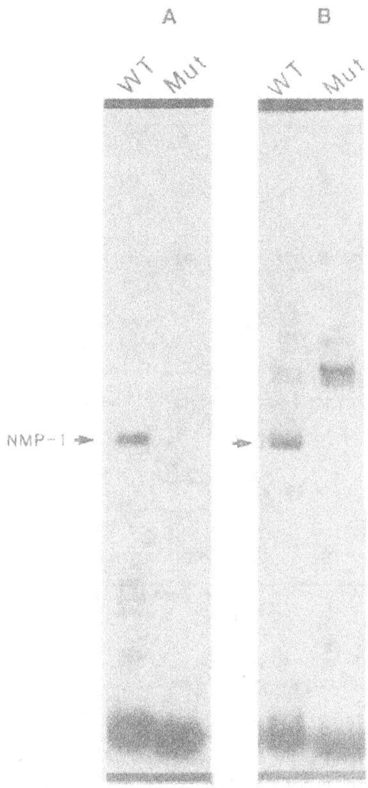

Fig. 10. (F) Nuclear matrix versus transcription factor extracts. The NMP-1 interaction can be found in transcription factor extracts. The mobility of the shift DNA appears to be the same; although it is demonstrated to bind to the same DNA sequence by using site-directed mutagenesis. The mut lanes have the CG changed to GA within the CGTCA binding sequence. It is clearly shown that the NMP-1 interaction is completely abolished when using these probes with both extracts.

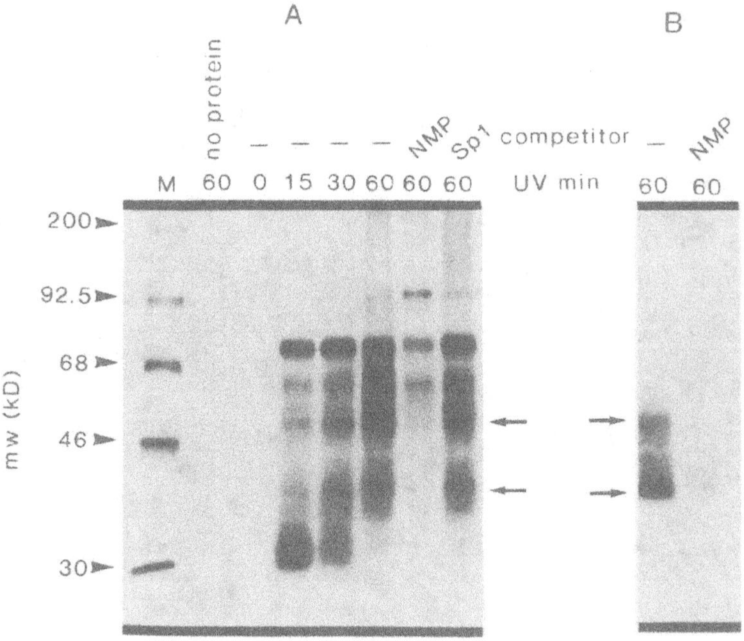

Fig. 10. (G) UV crosslinking experiments to determine the molecular weight of NMP-1. Panel A represents the UV crosslinking of total nuclear matrix extracts with the distal element and panel B shows the UV crosslinking the NMP oligonucleotide affinity chromatography fraction. The lower two bands in panel A (arrows) are the only bands competed out by the NMP oligonucleotide. Interestingly, only these bands appear when purified fractions are used (panel B, arrows). These proteins represent proteins of 43 and 54 kd molecular weight.

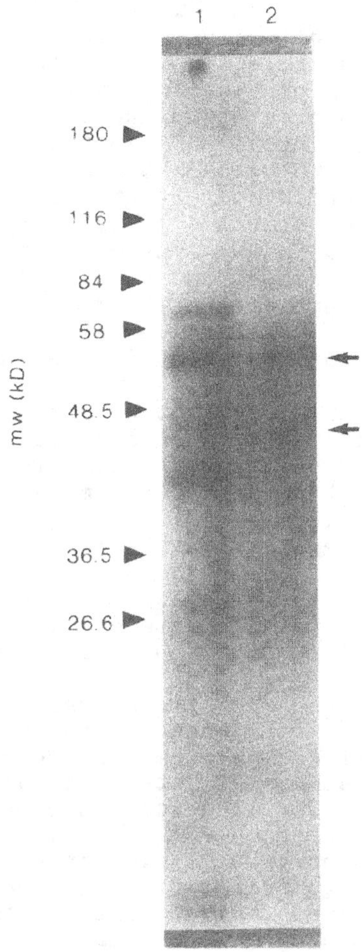

Fig. 10. (H) NMP-1 oligonucleotide affinity chromatography. Nuclear matrix extracts were passed over a Sepharose column that had the NMP oligonucleotide covalently linked. The column fractions are assayed by gel mobility shift experiments. The input represents the fraction before the column and the eluate is the fraction that was bound under 100 mM salt conditions and eluted with high salt. It is shown that the protein NMP-1 can be partially purified by the column..

determining the extent to which a protein is synthesized. Also there is a requirement to modulate the stability properties of mRNAs to facilitate turnover when genes are downregulated.

The biosynthesis of histones, the major structural proteins of chromatin, is temporally as well as functionally coupled to DNA replication[57-61]. At the natural termination of S phase as well as following interruption of DNA synthesis, there is a rapid and selective reduction in histone protein synthesis with a concomitant and stoichiometric decrease in steady state levels of histone mRNA. Selective destabilization of histone mRNA in the absence of DNA replication appears to be mediated at the post-transcriptional level[62-68]. The kinetics and extent of histone mRNA degradation following inhibition of DNA synthesis are not affected by treatment of cells with RNA synthesis inhibitors and histone gene transcription is not reduced significantly following inhibition of DNA replication[69]. However, to date, mechanisms involved in histone mRNA turnover are minimal and those presently understood are reviewed by Peltz, et al[70].

A viable mechanism to account for histone mRNA turnover must incorporate the ability of nucleases to preferentially utilize histone mRNAs as substrates. This mechanism may involve the recognition of histone mRNAs by a histone specific nuclease, the activity or availability of which is modified as a function of DNA replication. Alternatively, the effectiveness of histone mRNAs as substrates for nucleases with broad specificity may be related to recognition of sequences or structural elements of histone mRNAs that become accessible when DNA replication is interrupted.

Regardless of the specific mechanism that is operative, the questions arise as to whether the subcellular location of histone mRNA-containing polysomes is functionally related to the coupling of histone mRNA stability with DNA replication and whether subcellular localization can provide for the sequestering and/or concentration of macromolecules involved in histone mRNA turnover. Consistent with such reasoning, results from _in vitro_ translation and nucleic acid hybridization analyses indicate that, although histone mRNAs are found predominantly on non-membrane bound polysomes[71,72], these histone mRNA- containing polysomes are associated with the cytoskeleton[73]. This association with the cytoskeleton may provide a structural basis for the localization of histone mRNAs in specific regions of the cytoplasm.

To address experimentally the contribution of subcellular location to histone mRNA stability, we constructed a chimeric gene encoding a cell cycle dependent human H3 histone protein which was fused to a beta lactamase signal peptide sequence. The rationale for this approach was to introduce into proliferating cells a gene encoding H3 histone mRNA that can associate with membrane-bound polysomes rather than free polysomes and thereby determine whether the presence of histone mRNA on non-membrane-bound polysomes is requisite for the coupling of histone mRNA stability with ongoing DNA synthesis.

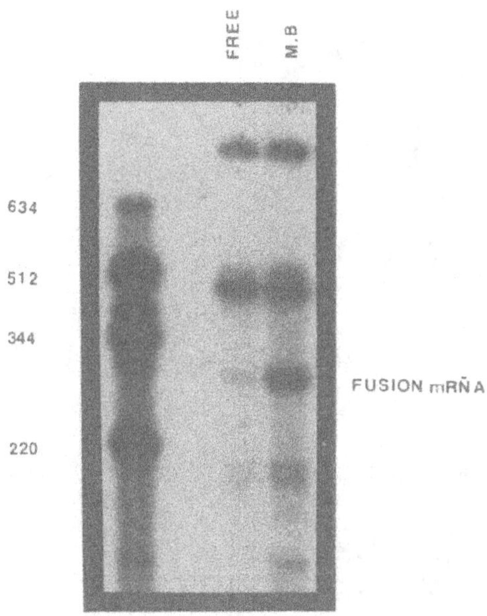

Fig. 11. *Association of signal peptide-histone fusion mRNA with membrane bound polysomes. Free polysomes and membrane bound polysomes were isolated from cells transfected with the chimeric gene encoding a signal peptide-histone fusion mRNA sequence. The RNAs isolated from both classes of polysomes were analyzed by S-1 nuclease protection analysis. The results indicate that approximately 30% of the signal peptide-histone fusion mRNA is found in the free polysome fraction while approximately 70% is associated with the membrane bound polysomes. In contrast, endogenous H3 histone mRNA is represented by greater than 90% in the free polysome fraction and less than 10% in the membrane bound polysome fraction.*

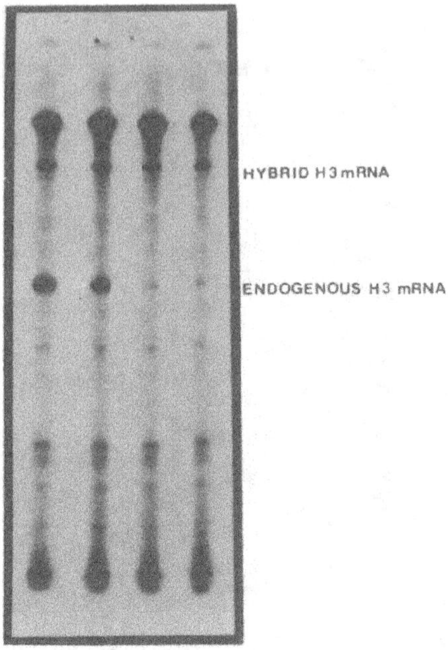

Fig. 12. *Stability of signal peptide-histone fusion mRNA following inhibition of DNA synthesis. 46 hours following transfection cells were treated with 1 mM hydroxyurea to inhibit DNA synthesis. Total cellular RNA from control and hydroxyurea-treated cells was analyzed by S-1 nuclease protection analysis to determine the stability of the endogenous (control) histone mRNA and the signal peptide-histone fusion mRNA. The left two lanes represent total cellular RNA from untreated cells and the right two lanes contain total cellular RNA from hydroxyurea-treated cells.*

The results presented in Figure 11 indicate that when the signal peptide histone fusion gene is transfected into actively proliferating cells expression occurs and the fusion message is targeted to the membrane-bound polysomes. As expected, the H3 mRNA transcribed from the endogenous histone gene is found associated with the non-membrane-bound polysomes. We then utilized the ability to direct a cell cycle dependent histone mRNA to membrane-bound polysomes to address the involvement of subcellular location in the coupling of histone mRNA stability with DNA replication. Actively proliferating cells were transfected with the signal peptide-histone fusion gene and as shown in Figure 12, following inhibition of DNA replication by hydroxyurea treatment, there is a rapid reduction (approximately 95%) in endogenous histone mRNA levels compared with those in untreated cells. In contrast, only a four per cent reduction in the signal peptide-histone fusion mRNA levels was observed. These results suggest that the stability properties of histone mRNA may be functionally related to subcellular location. The presence of histone mRNA on membrane-bound polysomes rather than on free polysomes where histone mRNA normally resides, may separate the message from the factor(s) that is involved in the selective destabilization of histone mRNA that occurs when DNA synthesis is inhibited. Such reasoning is further supported by in situ hybridization analysis which suggests that histone mRNA appears to be non-homogeneously distributed in the cytoplasm[74].

While these results fall short of providing the basis for a specific mechanism to account for the contribution of subcellular localization to mRNA stability there is support for a potentially important contribution of the cytoskeleton in modulating the cytoplasmic distribution of mRNAs as well as directly or indirectly modulating stability properties.

CONCLUSIONS

In summary, we have presented results that support a functional relationship between cell structure and the onset as well as progression of gene expression that results in the development of the osteoblast phenotype. This relationship involves both positive and negative control and is mediated at the transcriptional level as well as by modulating mRNA stability. Several lines of evidence have been presented which are consistent with active participation of nuclear and cytoarchitecture in the regulation of gene expression at multiple levels during the osteoblast developmental sequence, providing examples of how cell structure may directly contribute to both transcriptional control and mRNA stability.

ACKNOWLEDGEMENTS

Studies reported in this chapter were supported by grants from the National Institutes of Health (AR33920, AR35166, AR39588, GM32010, GM32381), the March of Dimes Birth Defect Foundation, the Northeast Osteogenesis Imperfecta Society and National Science Foundation (BMS 88-19989).

REFERENCES

1. G. Rovera, D. Santoli, and C. Damsky, Human promyelocytic leukemia cells in culture differentiate into macrophage-like cells when treated with a phorbol diester, Proc. Natl. Acad. Sci., USA, 76:2779-2783. (1979)
2. E. Huberman and M. Callahan, Induction of terminal differentiation in human promyelocytic leukemia cells by tumor-promoting agents, Proc. Natl. Acad. Sci. USA, 76:1293-1297 (1979).
3. H. Green, Adipose conversion: a program of differentiation, in "Obesity, Cellular and Molecular Aspects", G. Ailhaud, ed., Editions INSERM, Paris, pp. 15-25 (1979).
4. B. M. Spiegelman and H. Green, Control of specific protein biosynthesis during the adipose conversion of 3T3 cells, J. Biol. Chem., 255:8811-8818 (1980).
5. B. Nadal-Ginard, Commitment, fusion, and biochemical differentiation of a myogenic cell line in the absence of DNA synthesis, Cell, 15:855-864 (1978).
6. K. Okazaki and H. Holtzer, Myogenesis: fusion, myosin synthesis and the mitotic cycle, Proc. Natl. Acad. Sci. USA., 56:1484-1488 (1966).
7. J.G. Rheinwald and H. Green, Serial cultivation of strains of human epidermal keratinocytes: the formation of keratinizing colonies from single cells, Cell, 6:331-344 (1975).
8. R.H. Rice and H. Green, Presence in human epidermal cells of a soluble protein precursor of the cross-linked envelope: activation of the cross-linking by calcium ions, Cell, 18:681-694 (1979).
9. C.G. Bellows, J.E. Aubin, J.N.M. Heersche, and M.E. Antosa, Mineralized bone nodules formed in vitro from enzymatically released rat calvaria populations, Calcif. Tissue Int., 38:143-154 (1986).
10. U. Bhargava, M. Bar-Lev, C.G. Bellows and V. Aubin, Ultrastructural analysis of bone nodules formed in vitro by isolated fetal rat calvarial cells, Bone, 9:155-163 (1988).
11. B. Escarot-Charrier, F.H. Glorieux, M. van der Rest and G. Pereira, Osteoblasts isolated from mouse calvaria initiate matrix mineralization in culture, J. Cell Biol., 96:639-643 (1983).
12. L.C. Gerstenfeld, S.D. Chipman, J. Glowacki and J.B. Lian, Expression of differentiated function by mineralizing cultures of chicken osteoblasts, Dev. Biol., 122:49-60 (1987).
13. M.A. Aronow, L.C. Gerstenfeld, T.A. Owen, M.S.Tassinari, G.S. Stein and J.B. Lian, Factors that promote progressive development of the osteoblast phenotype in cultured fetal rat calvaria cells, J. Cell. Physiol., 143:213-221 (1990).
14. T.A. Owen, M. Aronow, V. Shalhoub, L.M. Barone, L. Wilming, M.S. Tassinari, M.B. Kennedy, S. Pockwinse, J.B. Lian and G.S. Stein, Progressive development of the rat osteoblast phenotype in vitro: reciprocal relationships in expression of genes associated with osteoblast proliferation and differentiation during formation of the bone extracellular matrix, J. Cell. Physiol., 143:420-430 (1990).

15. B. Escarot-Charrier, N. Shepard, G. Charette, M. Grynpas and F.H. Glorieux, Mineralization in osteoblast cultures: a light and electron microscopic study, Bone, 9:147-154 (1988).
16. G.S. Stein, J.B. Lian, L. Gerstenfeld, V. Shalhoub, M. Aronow, T. Owen and E. Markose, The onset and progression of osteoblast differentiation is functionally related to cellular proliferation, Conn. Tissue Res., 20:3-13 (1989).
17. M.J. Glimcher, Mechanism of calcification in bone: role of collagen fibrils and collagen phosphoprotein complexes in vitro and in vivo, Anatomical Record, 224:139-153 (1989).
18. K. Yoon, R. Buenaga and G.A. Rodan, Tissue specificity and developmental expression of osteopontin, Biochem. Biophys. Res. Comm., 148:1129-1136 (1987).
19. J.L. Carrington, A.B. Roberts, K.C. Flanders, N.S. Roche and A.H. Reddi, Accumulation, localization, and compartmentation of transforming growth factor β during endochondral bone development, J. Cell Biol., 107:1969-1975 (1988).
20. R. Bortell, L.M. Barone, M.S. Tassinari, J.B. Lian and G.S. Stein, Coordinate expression of transforming growth factor β_1 and collagen type I genes, but not collagen type II, during endochondral bone formation, J. Cell Biochem., 44:81-91 (1990).
21. M. Centrella, T.L. McCarthy and E. Canalis, Skeletal tissue and transforming growth factor β, FASEB J., 2:3066-3073 (1988).
22. U. Rüther, C. Garber, D. Komitowski, Müller and E.F. Wagner, Deregulated c-fos expression interferes with normal bone development in transgenic mice, Nature, 325:412-416 (1987).
23. S. Nomura, A.J. Wills, D.R. Edwards, J.K. Heath and B.L.M. Hogan, Developmental expression of 2ar (osteopontin) and SPARC (osteonectin) RNA as revealed by in situ hybridization, J. Cell Biol., 106:441-450 (1988).
24. A. Oldberg, A. Franzen and D. Heinegard, Cloning and sequence analysis of rat bone sialoprotein (osteopontin) cDNA reveals an Arg-Gly-Asp cell-binding sequence, Proc. Natl. Acad. Sci. USA., 83:8819-8823 (1986).
25. D. Heinegård and Å. Oldberg, Structure and biology of cartilage and bone matrix noncollagenous macromolecules. FASEB J., 3:2042-2051 (1989).
26. J.B. Lian and P.A. Friedman, The vitamin K-dependent synthesis of γ-carboxyglutamic acid by bone microsomes J. Biol. Chem., 253:6623-6626 (1978).
27. J.B. Lian, M.C. Coutts and E. Canalis, Studies of hormonal regulation of osteocalcin synthesis in cultured fetal rat calvariae, J. Biol. Chem., 260:8706-8710 (1985).
28. A.M. Craig, J.H. Smith and D.T. Denhardt, Osteopontin, a transformation-associated cell adhesion phosphoprotein, is induced by 12-0-tetradecanoylphorbol 13-acetate in mouse epidermis, J. Biol. Chem., 264: 9682-9689 (1989).
29. M.J. Glimcher, Role of collagen and phosphoproteins in the calcification of bone and other collagenous tissues, in "Calcium in Biological Systems". R.P. Rubin, G.B. Weiss, and J.W. Putney, eds. Plenum Press, New York. pp. 607-616 (1985).

30. J.B. Lian and C.M. Gundberg, Clinical significance of serum osteocalcin and urine Gla measurements, Clin. Orthop. Relat. Res., 266:267-291 (1988).
31. R.J. Majeska, B.C. Nair and G.A. Rodan, Glucocorticoid regulation of alkaline phosphatase in the osteoblastic osteosarcoma cell line ROS 17/2.8, Endocrinology, 116:1,170-179 (1985).
32. M. Plumb, J. Stein and G. Stein, Coordinate regulation of multiple histone mRNAs during the cell cycle in HeLa cells, Nucleic Acids Res., 11:2391-2410 (1983).
33. L.L. Baumbach, G.S. Stein and J.L. Stein, Regulation of human histone gene expression: transcriptional and posttranscriptional control in the coupling of histone messenger RNA stability with DNA replication, Biochemistry, 26:6178-6187 (1987).
34. T.A. Owen, J. Holthuis, E. Markose, A.J. van Wijnen, J.B. Lian and G.S. Stein, Modifications of protein-DNA interactions in the proximal promoter of a cell growth regulated histone gene during the onset of osteoblast differentiation. Proc. Natl. Acad. Sci., 87:5129-5133 (1990).
35. U. Pauli, S. Chrysolgelos, G. Stein, J. Stein and H. Nick, Protein-DNA interactions in vivo upstream of a cell cycle-regulated human H4 histone gene, Science 236:1308-1311 (1987).
36. A. van Wijnen, K.L. Wright, J.B. Lian, J.L. Stein and G.S. Stein, Human H4 histone gene transcription requires the proliferation-specific nuclear factor HiNF-D, J. Biol. Chem., 264:15034-15042 (1989).
37. V. Shalhoub, L.C. Gerstenfeld, D. Collart, J.B. Lian and G.S. Stein, Down-regulation of cell growth and cell cycle regulated genes during chick osteoblast differentiation with the reciprocal expression of histone gene variants, Biochemistry, 28:5318-5322 (1989).
38. G. Stein, J. Lian, J. Stein, R. Briggs, V. Shalhoub, K. Wright, U. Pauli and A. van Wijnen, Altered binding of human histone gene transcription factors during the shutdown of proliferation and onset of differentiation in HL-60 cells., Proc. Natl. Acad. Sci., USA, 86:1865-1869 (1989).
39. A. Lichtler, M.L. Stover, J. Angilly, B. Kream and D.W. Rowe, Isolation and characterization of the rat $\alpha 1(I)$ collagen promoter, J. Biol. Chem., 264:3072-3077 (1989).
40. E.R. Markose, J.L. Stein, G.S. Stein and J.B. Lian, Vitamin D-mediated modifications in protein-DNA interactions at two promoter elements of the osteocalcin gene, Proc. Natl. Acad. Sci. USA., 87:1701-1705 (1990).
41. T.A. Owen, R. Bortell, S.A. Yocum, S.L. Smock, M. Zhang, V. Shalhoub, N. Aronin, K.L. Wright, A.J. van Wijnen, C. Abate, J.L. Stein, T. Curran, J.B. Lian and G.S. Stein, Phenotype suppression of osteocalcin gene expression in osteoblasts by coordinate occupancy of the AP-1 sites in the vitamin D responsive and CAAT box elements by Fos-Jun. Proc. Natl. Acad. Sci., U.S.A., In Press.
42. S.B. Rodan, G. Wesolowski, K. Yoon and G.A. Rodan, Opposing effects of fibroblast growth factor and pertussis toxin on alkaline phosphatase, osteopontin, osteocalcin, and type I collagen mRNA levels in ROS 17/2.8 cells, J. Biol. Chem., 264:19934-19941 (1989).

43. M. Noda and G.A. Rodan, Type β transforming growth factor (TGFβ) regulation of alkaline phosphatase expression and other phenotype-related mRNAs in osteoblastic rat osteosarcoma cells, J. Cell. Physiol., 133:426-437 (1987).
44. J. Holthuis, T.A. Owen, A.J. van Wijnen, K.L. Wright, A. Ramsey-Ewing, M.B. Kennedy, R. Carter, S.C. Cosenza, K.J. Soprano, J.B. Lian, J.L. Stein and G.S. Stein, Tumor cells exhibit deregulation of the cell cycle histone gene promoter factor HiNF-D, Science, 247:1454-1457 (1990).
45. E.G. Fey and S. Penman, Nuclear matrix proteins reflect cell type of origin in cultured human cells, Proc. Natl. Acad. Sci., 85:121-5 (1988).
46. D.M. Pardoll, B. Vogelstein and D. Coffey, A fixed site of DNA replication in eukaryotic cells, Cell, 19:527-536 (1980).
47. B. Nelkin, D. Pardoll and B. Vogelstein, Localization of SV40 genes within supercoiled 100P domains, Nucl. Acids Res. 8:5623-5633 (1980).
48. S.I. Robinson, B.D. Nelkin and B. Vogelstein, The ovalbumin gene is associated with the nuclear matrix of chicken oviduct cells, Cell, 28:99-106 (1982).
49. C.A.G. van Eekelen and W.J. van Venrooij, hnRNA and its attachment to a nuclear protein matrix, J. Cell Biol., 88:554-563 (1981).
50. S. Zeitlin, A. Parent, S. Silverstein and A. Efstratiadis, Pre-mRNA splicing and the nuclear matrix, Mol. Cell. Biol., 7:111-120 (1987).
51. E. Barrack and D. Coffey, In Gene Regulation by Steroid Hormones, II, (A.K. Roy and J.H. Clark, eds.) Springer-Verlig, New York, p. 239-266 (1983).
52. M.H. Kumara, L.H. Shapiro and M.I. Surks, Association of the 3,5,3'-triiodo-L-thyronine nuclear receptor with the nuclear matrix of cultured growth hormone-producing rat pituitary tumor cells, J. Biol. Chem., 261:2844-2852 (1986).
53. D.G. Capco, K.M. Wan and S. Penman, The nuclear matrix: three dimensional architecture and protein composition, Cell, 29:847-858 (1982).
54. E.G. Fey, K.M. Wan and S. Penman, Epithelial cytoskeletal framework and nuclear matrix-intermediate filament scaffold: three dimensional organization and protein composition, J. Cell Biol.. 98:1973-1984 (1984).
55. D. He, J. Nickerson and S. Penman, Core filaments of the nuclear matrix, J. Cell Biol., 110:569-580 (1990).
56. E.G. Fey, G. Krochmalnic and S. Penman, The nonchromatin substructures of the nucleus: the ribonucleoprotein (RNP)-containing and RNP-depleted matrices analyzed by sequential fractionation and resinless section electron microscopy, J. Cell Biol., 102:1654-1665 (1986).
57. F. Marashi, L. Baumbach, R. Rickles, F. Sierra, J.L. Stein and G.S. Stein, Histone proteins in HeLa S3 cells are synthesized in a cell cycle state specific manner. Science, 215:683-685 (1982).
58. J. Spaulding, K. Kajiwara and G.C. Mueller, The metabolism of basic proteins in HeLa cell nuclei, Proc. Natl. Acad. Sci. USA, 56:1535-1542 (1966).
59. E. Robbins and T.W. Borun, The cytoplasmic synthesis of histones in HeLa cells and its temporal relationship to

DNA replication, Proc. Natl. Acad. Sci. USA, 57:409-416 (1967).
60. G.S. Stein and T.W. Borun, The synthesis of acidic chromosomal proteins during the cell cycle of HeLa S3 cells: I. The accelerated accumulation of acidic residual nuclear protein before the initiation of DNA replication, J. Cell Biol., 52:292-307 (1972).
61. R.S. Wu and W.M. Bonner, Separation of basal histone synthesis from S-phase histone synthesis in dividing cells, Cell, 27:321-330 (1981).
62. W.B. Butler and G.C. Mueller, Control of histone synthesis in HeLa cells, Biochim. Biophys. Acta, 294:481-496 (1973).
63. S. Helms, L. Baumbach, G. Stein and J. Stein, Requirement of protein synthesis for the coupling of histone mRNA levels and DNA replication, FEBS Lett., 168:65-69 (1984).
64. L. Baumbach, F. Marashi, M. Plumb, G. Stein and J. Stein, Inhibition of DNA replication coordinately reduces cellular levels of core and H1 histone mRNAs" requirement for protein synthesis, Biochemistry, 23:1618-1625 (1984).
65. D. Sittman, R.A. Graves and W.F. Marzluff, Histone mRNA concentrations are regulated at the level of transcription and mRNA degradation, Proc. Natl. Acad. Sci. USA, 80:1849-1853 (1983).
66. N. Heintz, H.L. Sive and R.G. Roeder, Regulation of human histone gene expression: kinetics of accumulation and changes in the rate of synthesis and in the half-lives of individual histone mRNAs during the HeLa cell cycle, Mol. Cell. Biol., 3:539-550 (1983).
67. M. Plumb, J. Stein and G. Stein, Coordinate regulation of multiple histone mRNAs during the cell cycle in HeLa cells, Nucleic Acids Res., 11:2391-2410 (1983).
68. M. Plumb, J. Stein and G. Stein, Influence of DNA synthesis inhibition on the coordinate expression of core human histone genes during S phase, Nucleic Acids Res., 11:7927-7945 (1983).
69. L. Baumbach, J. Stein and G. Stein, Regulation of human histone gene expression: transcriptional and posttranscriptional control in the coupling of histone messenger RNA stability with DNA replication, Biochemistry, 26:6178-6187 (1987).
70. S.W. Peltz, G. Brewer, P. Bernstein, P.A. Hart and J. Ross, Regulation of mRNA Turnover in Eukaryotic Cells, Crit. Rev. in Eukaryotic Gene Exp., in press, (1990).
71. J.L. Stein, C.L. Thrall, W.D. Park, R.J. Mans and G.S. Stein, Hybridization analysis of histone messenger RNA: Association with polyribosomes during the cell cycle, Science, 189:557-558 (1975).
72. T.W. Borun, M.D. Scharff and E. Robbins, Rapidly labeled polyribosome-associated RNA having the properties of histone messenger, Proc. Natl. Acad. Sci. USA, 58:1977-1983 (1967).
73. G. Zambetti, W. Schmidt, G. Stein and J. Stein, Subcellular localization of histone messenger RNAs on cytoskeleton-Associated free polysomes in HeLa S3 cells, J. Cell. Physiol., 125:345-353 (1985).
74. J.B. Lawrence, R.H. Singer, C.A. Villnave, J.L. Stein and G.S. Stein, Intracellular distribution of histone mRNAs in human fibroblasts studied by in situ hybridization, Proc. Natl. Acad. Sci, USA, 85:463-467 (1988).

CELL STRUCTURE AND GENE EXPRESSION: CONTRIBUTIONS OF THE EXTRACELLULAR MATRIX TO REGULATION OF OSTEOBLAST GROWTH AND DIFFERENTIATION

Jane B. Lian, Gary S. Stein, Thomas A. Owen, Michael Aronow, Melissa S. Tassinari, Shirwin Pockwinse, Rita Bortell

Department of Cell Biology, University of Massachusetts Medical center, 55 Lake Avenue North, Worcester, MA 01655

INTRODUCTION

The role of the extracellular matrix of specialized tissues in promoting cellular differentiation has long been recognized (1-2). Our studies have utilized rat osteoblast cultures (3-5) as a model system to examine a well defined extracellular matrix (ECM) and the coordinate regulation of the changes in cell structure and gene expression as related to its formation. These cells produce a mineralized ECM having a bone tissue-like organization analogous to embryonic bone (4-6). In contrast to tumor-derived or transformed osteosarcoma cell lines (7), those normal diploid bone derived cells in culture exhibit normal cell cycle regulated expression of genes which is functionally coupled to expression of differentiation specific genes (4). As the ECM develops, osteoblasts differentiate, progressing through three stages of development, a proliferation period, a period of matrix maturation, and a mineralization period (4). In the previous chapter, temporal expression of genes characterizing the osteoblast phenotype associated with this differentiation in vitro has been described in detail and is summarized in Figure 1.

Initially, a matrix composed primarily of collagen and fibronectin forms during the proliferative period and then "matures." Two transition points occur during this sequence. The first occurs with the shutdown of proliferation and is causally related to induction of genes (4), such as alkaline phosphatase, that are required for modifications of the ECM which lead to mineralization and further differentiation of the cells. The second transition point is at the onset of mineralization at which time a third subset of genes become induced, some unique to the osteoblast phenotype, such as osteocalcin (Figure 1).

Molecular Basis of Human Cancer
Edited by C. Nicolini, Plenum Press, New York, 1991

Figure 1. Development of the Osteoblast Phenotype.
(A) Light microscopy illustrating 3 periods of osteoblast differentiation. *Panel 1 (Day 11) shows ~ 80% of the cells proliferating (3H thymidine autoradiography). Cells will multilayer in nodules where proliferation shuts down and initial expression of alkaline phosphatase activity is visualized. Panel 2 (Day 21) represents osteoblasts stained for the enzyme alkaline phosphatase. Panel 3 (Day 35) shows the mineralized nodules after the von Kossa silver stain.*

(B) Temporal expression of genes that define three principal periods of the osteoblast developmental sequence (proliferation, matrix development and maturation, mineralization). *The functional relationship between proliferation and differentiation is schematically illustrated by the arrows within the context of down-regulation of proliferation associated genes (H4 histone and AP-1 binding activity) and the up-regulation of genes associated with the maturation (type αI collagen, alkaline phosphatase) and mineralization (hydroxyapatite (HA) deposition, osteopontin (OP), and osteocalcin (OC)) of the osteoblast ECM.*

(C) Illustration of the Role of the Extracellular Matrix (ECM) in Promoting Osteoblast Growth and Differentiation. *The proliferation period supports the synthesis of a type I collagen/fibronectin ECM which continues to mature and mineralize. The formation of this matrix down-regulates proliferation and matrix mineralization down-regulates the expression of genes associated with the formation-maturation period. Also indicated is a model for suppression of two marker genes of the mature osteoblast phenotype (alkaline phosphatase and osteocalcin) in actively proliferating cells by protein binding to AP-1 sites. The occupancy of the AP-1 sites in the alkaline phosphatase and osteocalcin gene promoter by Fos-Jun and/or related proteins suppresses the basal and vitamin D induced expression of the osteocalcin gene prior to the initiation of alkaline phosphatase and osteocalcin basal expression following the proliferative period. AP-1, AP-1 binding activity; H4, H4 histone; COLL-αI, Type αI collagen; ALK PHOS, alkaline phosphatase; OP, osteopontin; OC, osteocalcin; HA, total accumulate hydroxyapatite (calcium + phosphate).*

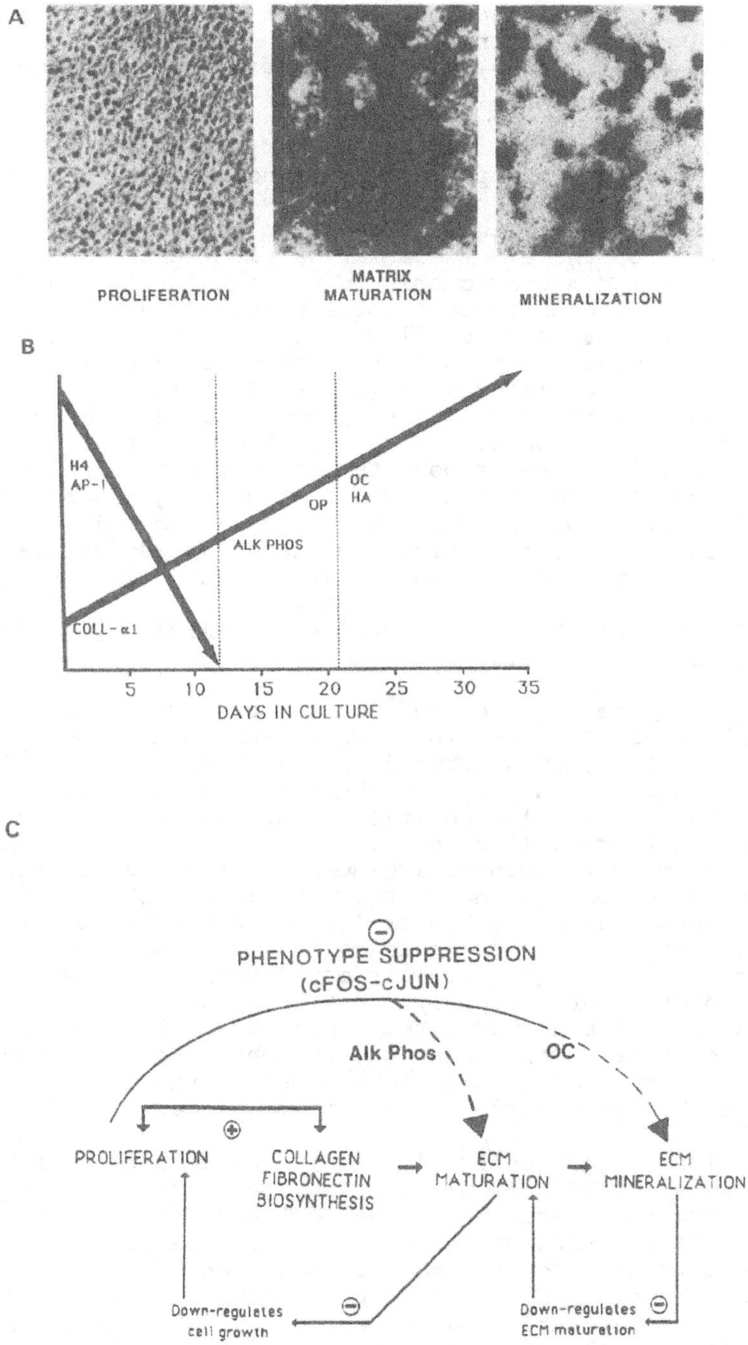
A
PROLIFERATION
MATRIX
MATURATION
MINERALIZATION
B
H4
AP-1
COLL-α1
ALK PHOS
OP
OC
HA
5
10
15
20
25
30
35
DAYS IN CULTURE
C
PHENOTYPE SUPPRESSION
(cFOS-cJUN)
Alk Phos
OC
PROLIFERATION
COLLAGEN
FIBRONECTIN
BIOSYNTHESIS
ECM
MATURATION
ECM
MINERALIZATION
Down-regulates
cell growth
Down-regulates
ECM maturation

Proteins expressed with the onset of mineralization or within this last period likely function in the continued formation of the mineralized matrix as well as in the maintenance and turnover of bone, the latter being a matter needing further investigation. Osteocalcin, for example, has been shown to function in stimulation of osteoclastic activity for bone resorption (8-9) and osteopontin is associated with the osteoclast membrane (10).

These transition points are so indicated since osteoblast differentiation can proceed up to but not pass beyond these points until further signals are received. Recent evidence indicates that both of these transition points depend upon modifications of the extracellular matrix (ECM) for the expression of genes characteristic of that period and the ECM can in turn feed back or contribute to the down-regulation of genes in a previous period (Figure 1C). It is now understood, albeit incompletely, that proto-oncogene proteins (11) expressed during proliferation (4) can contribute to the suppression of certain phenotypic genes expressed only in a later period (12). We will also show the effect of physiologic modulators of bone formation for example the hormone 1,25 dihydroxyvitamin D3, in modifying the ECM thereby either disrupting or enhancing differentiation. Additionally, we will discuss hormonal effects of gene expression that are dependent upon the maturational state of the cell.

FORMATION OF THE BONE-LIKE EXTRACELLULAR MATRIX _IN VITRO_ REGULATES GENE EXPRESSION

In vivo, the major components of bone are the mineral phase, hydroxyapatite ($Ca6(OH)_{10}PO_4$) ranging from 70-80% of the total bone matrix and an organic phase consisting largely of type I collagen accounting for the majority (85-90%) of ECM protein (13). Mineral deposition occurs in association with the collagen fibrils (Figure 2A). Of the organic matrix, bone contains from 10-20% non-collagenous proteins, some uniquely accumulated in bone, although they are also synthesized in other tissues (Table I). Several studies suggest the bone phosphoproteins may function in early events of mineral formation (13) while osteocalcin may function in the regulation of mineral deposition or bone turnover by osteoclasts, the multinucleated resorbing cells of bone (8-9). Notably, both osteocalcin and one phosphoprotein, osteopontin (SPPI) are expressed in high levels during the mineralization period of the osteoblast development sequence.

In vitro, collagen is also the major component of the extracellular matrix (Figure 2) of osteoblast cultures accumulating throughout the 30 day culture period to account for as much as 30-50% of the extracellular matrix (3,14). These amounts are proportional to that found in developing fetal bone tissue (15). Collagen and fibronectin are produced during the proliferation period when their mRNA levels are the highest (Figure 1A); however, synthesis and accumulation continues throughout the culture period. Collagen in fact is processed more rapidly and accumulates more efficiently after the proliferation period (16).

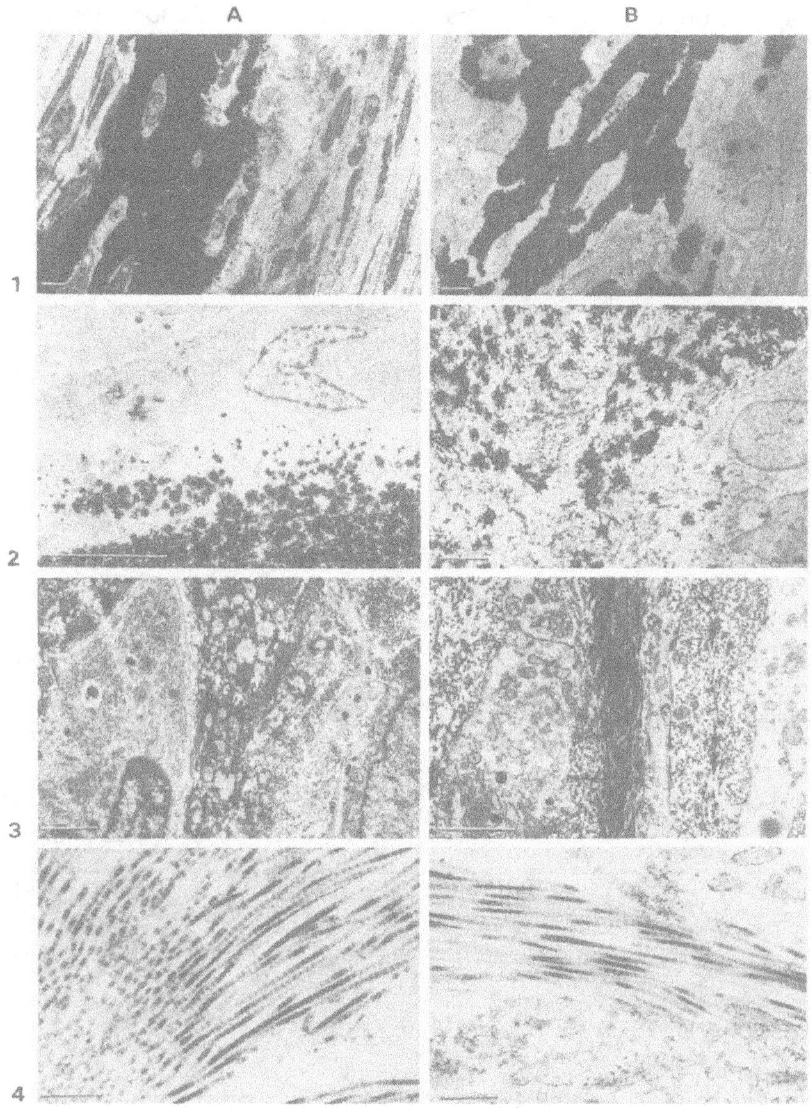

Figure 2. Transmission electron micrographs comparing cross sections of a 21 day fetal rat calvarium (column A) with those of a mineralized nodule formed in a 35 day osteoblast culture (column B). *Row 1 shows general tissue organization of cells and mineralized bone extracellular matrix enveloping the osteoblasts (arrow designates periosteal fibroblasts). Row 2 shows early stages of mineral deposition associated with collagen fibrils and the absence of intracellular calcification. Row 3 shows identical orthogonal organization of the collagen matrix in both the intact calvarium and in the mineralized nodule formed* ***in vitro****. For this micrograph in column A, a demineralized calvarium was sectioned. Row 4 shows higher magnification of the collagen bundles with mineral deposition within the collagen fibrils. Bars in the lower micrographs represent 2.5 μm; others represent 5 μm.*

Table 1 Non-collagenous Protein Products of Bone Cells

Constituent	Other names used	Molecular weight (mass, SDS gels)	References
Osteocalcin	Bone Gla protein (BGP), 49 residues	(10 kDa) 5.7 kDa	Hauschka et al Physiol Rev 1989 69:990
Osteopontin	Bone sialoprotein I (SPPI), 44 kDa phosphoprotein	(66 kDa)	Smith & Denhardt J Cell Biochem 1987 34:13
Osteonectin	SPARC, BM-40, 43 kDa protein	(45 kDa)	
Proteoglycan I	PG-1, Biglycan	(240 kDa)	Fisher et al 1987 J Biol Chem 262:9702
Proteoglycan II	PG-II, Decorin proteodermatan sulphate	(120 kDa)	Kinne et al 1987 J Biol Chem 262:10206
Bone sialoprotein	Sialoprotein II	(80 kDa)	Franzén & Heingård 1985 Biochem J 232:715
Matrix Gla[a] protein	MGP, 79 residue	10 kDa	Price 1988 Ann Rev Nut 8:865
24K phosphoprotein	α1 (I) Procollagen amino propetide[b]	24 kDa	Uchiyama et al 1986 Biochem 25: 7572
Growth factors	FGF,TGFβ,PPGF,IFI-I		Canalis et al 1989 J Endocrin Inv 12:577
Bone morphogenic proteins	BMP 1, BMP 2A, BMP 2B, BMP 3,	730 a.a (BMP-1) 396 (2A1)	Wozney et al 1988 Science 242:1528
	Osteogenin	28-43Kd	Sampath et al 1987 Proc Natl Acad Sci USA 84:7109

[a] Gla residue, γ-carboxyglutamic acid

Formation of the collagenous matrix contributes to changes in cell structure, osteoblast differentiation and gene expression in several ways. First, with formation of the collagen matrix changes in cell shape are observed, with the cells becoming markedly cuboidal as the collagen matrix surrounds these cells (3,5-6,14,17). The cells do not form a contact inhibited monolayer but have the ability to pile on top of each other forming a nodule multi-layers of cells surrounded by collagen matrix (3-6,14). _In vivo_, as the extracellular matrix mineralizes, the osteoblasts become more rounded and develop long processes for communication and, as osteocytes, provide nutriation to the mineralized matrix (18). Secondly, formation of the collagenous matrix contributes to the shut-down of proliferation which is a requisite for the induction of certain genes characteristic of the osteoblast phenotype. This relationship is clearly evident from the study in which cells were cultured with various concentrations

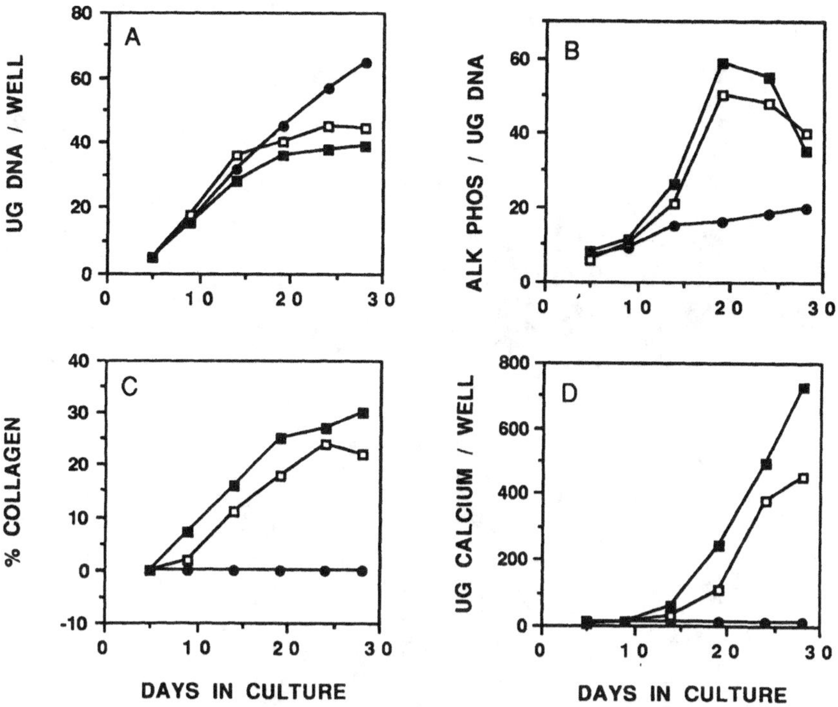

Figure 3. ***Effect of ascorbic acid and collagen accumulation on osteoblast proliferation (cell number) and differentiation parameters (collagen content, alkaline phosphatase activity and calcium deposition).*** *Cells were maintained during a 30 day culture period, first in MEM medium (until day 7) and then in BGJb medium supplemented with 10% FCS, 10 mM β-glycerol phosphate and containing 0 (●), 25 (□), or 50 (■) μg/ml ascorbic acid. Values plotted represent the mean of three sample wells per time point. (A) DNA was determined by a fluorometric assay. (B) Alkaline phosphatase is expressed as nM p-nitrophenol (released from the substrate p-nitrophenol phosphate) per microgram DNA. (C) Percent collagen accumulated in the extracellular matrix was determined from the hydroxyproline concentration in total amino acid analysis of the cell layers. (D) Calcium was determined by atomic absorption spectroscopy in the same samples hydrolyzed for the amino acid analysis. With increasing ascorbic acid concentrations, the cells reached confluency at a lower density and there was an increase in collagen accumulation associated with a parallel increase in alkaline phosphatase activity. In the absence of a collagen matrix, no accumulation of calcium occurred, reflecting the absence of mineralization.*

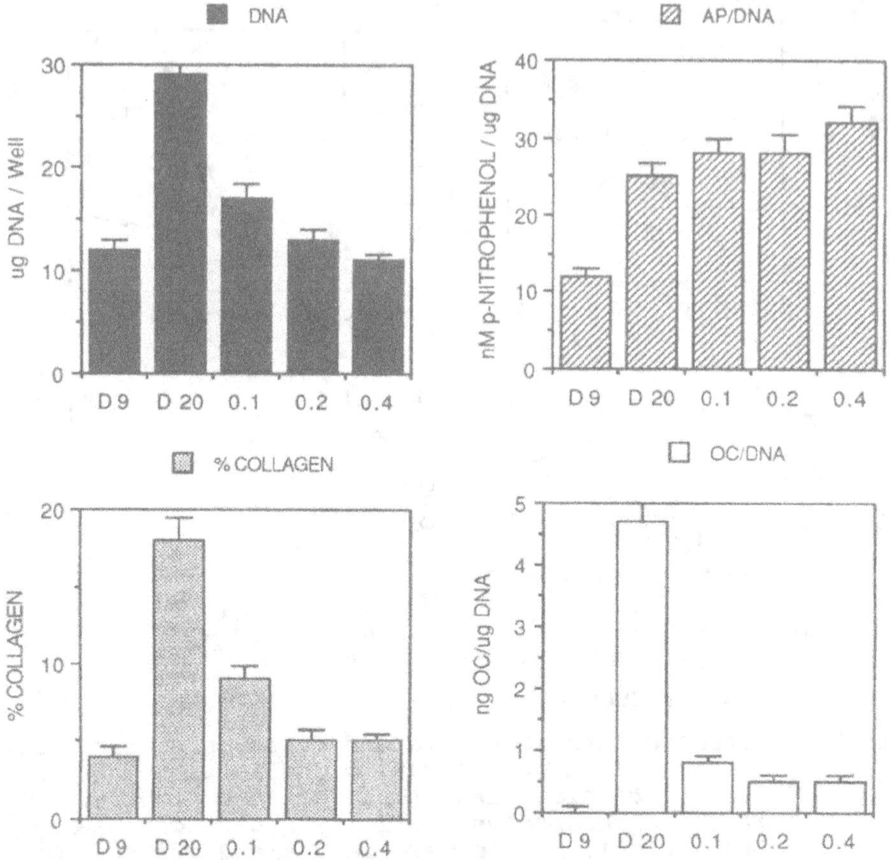

Figure 4. ***Hydroxyurea Treatment of Osteoblast Cultures.*** *Cultures were exposed to 0.1, 0.2 or 0.4 mM hydroxyurea (HU) from Day 9 (D9) to Day 20 (D20), at which time the cells were harvested for assays. D9 and D20 represent the two control untreated groups. Note from DNA, cell growth was inhibited without cell death up to 0.4 mM HU but some proliferation continued at the 0.1 mM dose. The inhibition of proliferation was coupled to a significant decrease in collagen synthesis measured as % collagen of total protein accumulated in the cell layers determined by amino acid analysis. Alkaline phosphatase activity (nm of product p-nitrophenol) increased from day 9 and was not significantly different from day 20 controls. Osteocalcin (OC) induced expression was inhibited as a result of the absence of mineralization in the HU treated cultures due to an insufficient formation of a collagen matrix. These changes support the concept of a functional relationship of proliferation to gene expression.*

of ascorbic acid, a requirement for collagen synthesis (Figure 3). In the absence of ascorbate and formation of a collagen containing ECM, the developmental sequence of osteoblast differentiation did not occur. In the presence of ascorbate, the higher the level of collagen synthesis and accumulation in the extracellular matrix, the lower the density at which cells cease to proliferate and, coordinately alkaline phosphatase enzyme mRNA levels and activity per cell are greater. Thus there is a contribution of signals from the extracellular matrix that promotes a higher level of gene expression than in the absence of the extracellular matrix.

Requirement for Proliferation and Collagen Synthesis

In the previous chapter, the functional relationship between the shut-down of proliferation and the onset of alkaline phosphatase gene expression was demonstrated by using hydroxyurea to inhibit proliferation and cause premature induction of a higher level of gene expression. However, the shut-down of proliferation alone is insufficient to promote the complete differentiation of osteoblasts. This is demonstrated in Figure 4 which shows the effects of chronic hydroxyurea exposure of the cultures at doses low enough to inhibit proliferation but maintain viability of the cells. With inhibition of proliferation, collagen mRNA production was decreased; less extracellular matrix was accumulated, which was insufficient to support mineralization or the expression of genes related to the mineralization period. An increase in alkaline phosphatase activity from the control starting point was observed similar to acute effects of hydroxyurea on mRNA levels, however, the enzyme activity levels were similar to the control cultures at the end point (20 days). This demonstrates again the requirement for formation of an extracellular matrix capable of mineralization to support progressive differentiation of the osteoblast.

Requirement for ECM Mineralization for Complete Osteoblast Differentiation

From these experiments (Figures 3 and 4), it was realized that a second transition in the developmental sequence occurs with mineral deposition. The onset of mineralization can be delayed by the removal of the β-glycerol phosphate from the culture medium as shown in Figure 5. The expression and induction of two calcium binding proteins, osteocalcin and osteopontin, were seen to be coordinately increased with the rate and extent of mineralization of the extracellular matrix as measured by calcium accumulation. This was not true for the calcium binding protein matrix Gla protein (data not shown). Mineralization of the extracellular matrix *in vitro* also modifies genes expressed in the previous period contributing to the downregulation of alkaline phosphatase enzyme activity and to the downregulation of genes expressed during the period of active mineral deposition; that is, osteocalcin and osteopontin become partially downregulated in heavily mineralized cultures (Figure 5). Such downregulation of genes in heavily mineralized cultures is analogous to *in vivo* activity of osteocytes, cells which are completely embedded within a mineralized matrix (19-20). In these cells

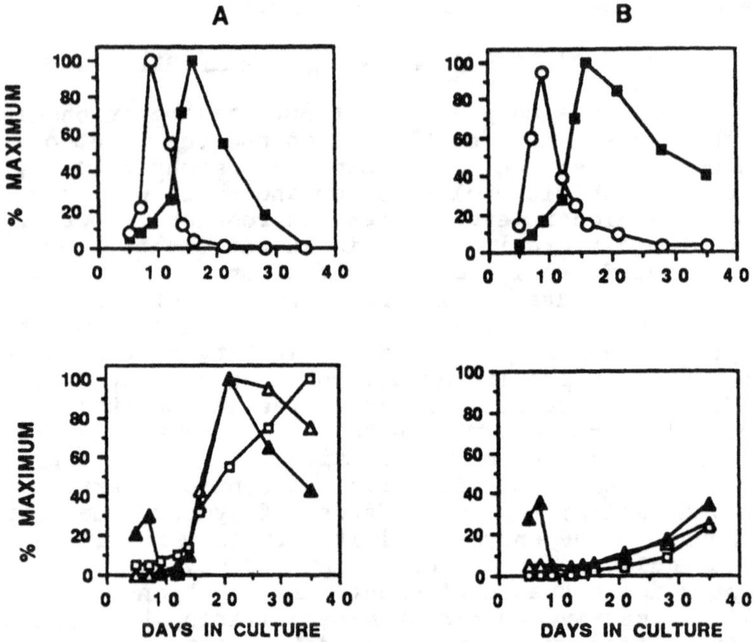

Figure 5. ***Relationship of mineral deposition to osteoblast gene expression.*** *Cultures maintained in BGJb medium supplemented with 50 μg/ml ascorbic acid and 10 mM β-glycerol phosphate (βGPO$_4$) (A) or in the absence of βGPO$_4$ which delays the onset and extent of calcification (B). Conditions that delay mineralization of the osteoblast cultures do not affect the earlier events of proliferation (^{3}H-thymidine incorporation -○-) and extracellular matrix maturation (collagen synthesis (not shown) and alkaline phosphatase activity -■-). In (B) alkaline phosphatase was not downregulated compared to (A) as a result of the delay in mineralization. The lower profiles show earlier accumulation of calcium (-□-) in the extracellular matrix and earlier osteocalcin (-△-) and osteopontin (-▲-) induced expression in (A) compared to cultures maintained in the same medium without β-glycerol phosphate (B).*

alkaline phosphatase activity and synthesis of extracellular matrix proteins is markedly diminished compared to cells in the mineralizing front where active formation and mineralization of the extracellular matrix is occurring.

MODIFICATIONS OF THE EXTRACELLULAR MATRIX AND OSTEOBLAST DIFFERENTIATION BY 1,25 DIHYDROXYVITAMIN D_3

Vitamin D in its active form, $1,25(OH)_2D_3$, is an important hormonal regulator of calcium homeostasis functioning primarily in the intestinal absorption of calcium and the mobilization of calcium from bone (21). This is accomplished via classic steroid hormone action with vitamin D binding to its cytosolic receptor for translocation to the nucleus where binding occurs to sequence specific elements in the gene. The vitamin D receptor complex modulates the expression of several proteins including the vitamin D dependent calcium binding protein, calbindin, in the intestine as well as several bone proteins that are involved in both formation, mineralization and turnover of the bone matrix (22). These include Type I collagen (23), the enzyme alkaline phosphatase (24) involved in early events of mineral deposition, osteopontin or bone sialoprotein I (25) and the 2 vitamin K dependent calcium binding proteins, osteocalcin (26-29) and matrix Gla protein (30). Vitamin D deficiency in vivo leads to rickets in children or osteomalacia in adults, in which there is decreased mineralization of the matrix and a decrease in the two vitamin D regulated proteins associated with accumulation of mineral, osteocalcin and osteopontin (31). Interestingly, alkaline phosphatase is elevated in vitamin D deficiency, consistent with observations of a downregulation of alkaline phosphatase mRNA and activity by vitamin D and in heavily mineralized bone tissue.

The normal diploid rat osteoblast cultures provide an ideal system to examine the effects of vitamin D on expression of bone parameters as a function of the differentiated state of the cell. At selected time points throughout a 34-day culture period 10^{-8}M $1,25(OH)_2D_3$ was added for 48 hours, an acute exposure. As shown in Figure 6, biphasic positive and negative regulation of three osteoblast parameters were observed, collagen type I, alkaline phosphatase, and histone H4 gene. In proliferating cells, when the mRNAs of these parameters were expressed at high levels, the hormone inhibited their expression. However, in the most differentiated cells within a mineralized matrix when these parameters were at their lowest basal levels, $1,25(OH)_2D_3$ increased mRNA levels. Such results suggest interactive signals between the basal transcription regulatory elements and the vitamin D responsive element which modulate expression. With three osteoblast phenotype markers (Figure 6) osteocalcin, osteopontin, and matrix gla protein (MGP), $1,25(OH)_2D_3$ stimulated mRNA levels in short term (48 hrs) exposure throughout the developmental sequence. It is of interest to note $1,25(OH)_2D_3$ did not induce osteocalcin expression in the proliferation period. This is not the result of the absence of 1,25 D_3 receptors since other vitamin D regulated genes, such as osteopontin MGP, are expressed and modulated in proliferating cells.

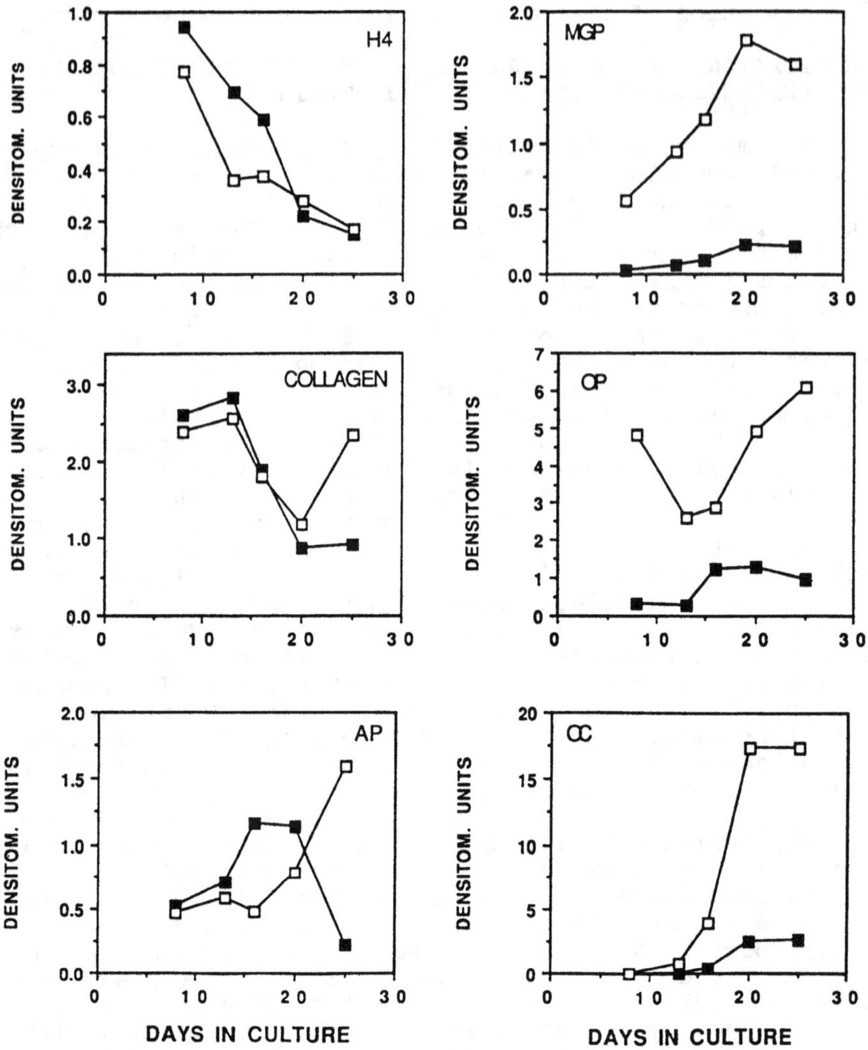

Figure 6. Cellular mRNA levels of rat osteoblast phenotype markers in control (■) and 48 hr. 10^{-8}M $1,25(OH)_2D_3$ treated cultures (-□-) during the developmental sequence of osteoblast differentiation. *Values are expressed for each parameter as % maximum value in control and vitamin D treated cultures for each gene. Each value is an average densitometric value for 2 Northern blot analyses of total cellular RNA prepared from cultures on Days 8, 13, 16, 20 and 25. Values were normalized to ribosomal RNA and the average values were within 7%. Parameters are H4 histone gene, a marker of DNA synthesis, collagen Type I, and alkaline phosphatase (ALK PHOS), matrix Gla protein (MGP), osteopontin (OP) and osteocalcin (OC). Note the biphasic regulation (inhibition, then stimulation) of collagen and AP by vitamin D.*

In contrast to these acute effects, a different profile of regulation emerged with chronic hormone treatments at 10^{-8}M $1,25(OH)_2D_3$ during the culture period. When the cultures were treated continuously 10^{-8}M $1,25(OH)_2D_3$ (chronic vitamin D treatment) from the proliferation period (Day 6) the hormone blocked further differentiation of the cells (Figures 7 and 8) and expression of parameters reflecting the mineralized differentiated state, such as formation of mineralized nodules (Figure 8) and osteocalcin and osteopontin expression. Only matrix Gla protein, expressed in the early proliferation, was not inhibited by chronic vitamin D, but stimulated at all time points. In cultures treated from Day 6-34, no mineralized nodules formed. This hormone-mediated restriction in the progression of osteoblast differentiation resulted from its downregulation of alkaline phosphatase and collagen mRNA production and the absence of collagen synthesis and formation of a mineralized extracellular matrix. Therefore, genes that are normally turned on with the onset of mineralization (e.g., osteocalcin) were blocked. This finding clearly demonstrates that the hormone can only modify ongoing basal levels of these expressed osteoblast genes and will not induce these genes to be turned on.

To further test the hypothesis that hormonal stimulation of these genes was dependent upon the formation of the bone extracellular matrix, chronic experiments were initiated during the second stage of maturation on day 20 after the initiation of formation of a mineralized matrix, and when AP levels were at there maximum. Mineralized nodules increased less in size compared to controls (Figure 8) and osteocalcin expression was no longer blocked as a consequence of the chronic treatment but was stimulated (Figure 7). However, osteopontin normally induced with matrix mineralization remained downregulated, indicating pleiotropic effects of $1,25(OH)_2D_3$ on this parameter was dependent upon duration of treatment (data not shown). Although these studies were carried out at high doses, such chronic effects may be important considerations in evaluating long term therapeutic use of $1,25(OH)_2D_3$ in the treatment of bone disorders.

$1,25(OH)_2D_3$ can promote differentiation of several progenitor cells including those which form the multinucleated osteoclasts, the resorbing cells of bone (32). Physiologic levels of vitamin D can both promote mobilization of calcium from bone, stimulate bone resorption and at the same time lead to replacement of the resorbed bone by formation of new bone tissue. Pharmacologic doses of vitamin D increase bone resorption and lead to hypercalcemia. These changes in formation and turnover of bone extracellular matrix modulated by vitamin D are largely a reflection of hormonal transcriptional regulation of osteoblast genes (22). Thus, it is clearly of interest to understand hormonal regulation of these osteoblast parameters as influenced by the differentiated state or microenvironment of the osteoblast.

The in vitro effects of vitamin D on some osteoblast parameters are variable, dependent upon the in vitro culture systems. Parameters such as collagen and alkaline phosphatase have been shown to be regulated by vitamin D biphasically,

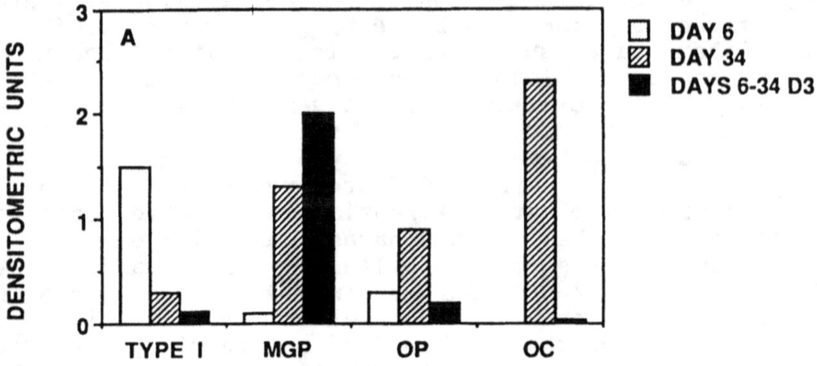

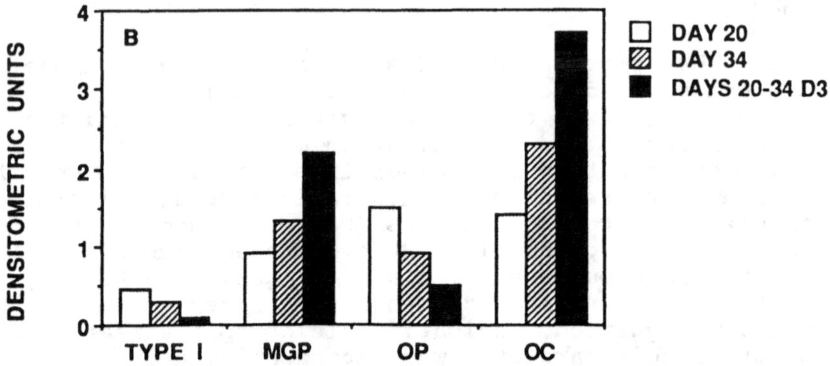

Figure 7. ***Chronic treatment of rat osteoblast culture initiated on Day 6 (top profile) during proliferation or on Day 20 (lower profile) with 10^{-8}M 1,25(OH)$_2$D$_3$.*** *Cellular mRNA levels of vitamin D regulated genes after chronic treatment of the rat osteoblast cultures with 10^{-8}M 1,25(OH)$_2$D$_3$ was initiated from Day 6 or from Day 20 of the developmental sequence. Two control values, the day of initiation of hormone at either Day 6 or Day 20 (grey bars) and the last day (Day 34, hatch bar) of the experiment are compared to the hormone treated cultures (solid bar). Note partial inhibition of osteopontin by chronic vitamin D initiated at either time and, for osteocalcin, the absence of expression on Day 6 and the trace level detected from Day 6 to 34 vitamin D-treated cultures in panel A compared to OC peak expression in Day 34 control. In contrast, OC is stimulated by vitamin D from Days 20-34 (panel B).*

dependent either on cellular density (33-34) or the source of osteoblast-like cells (34-36). Alkaline phosphatase, for example, will be stimulated by vitamin D when basal levels are low and will be inhibited by the hormone when basal levels are high (34). Vitamin D stimulates collagen synthesis in human osteoblast-like cell lines (35-36) but inhibits collagen synthesis in rat derived osteoblast cell lines (23,37). Thus, it appears that the very complex effects of vitamin D on osteoblast parameters is related to the phenotypic characteristics or differentiated state of the osteoblast-like cells in vitro. To understand these complex pleiotropic effects of $1,25(OH)_2D_3$ on gene expression requires a thorough analysis of the vitamin D responsive regulatory and primary transcriptional elements in these genes. Our laboratory is defining and characterizing the interactions between various regulatory elements in the osteocalcin gene.

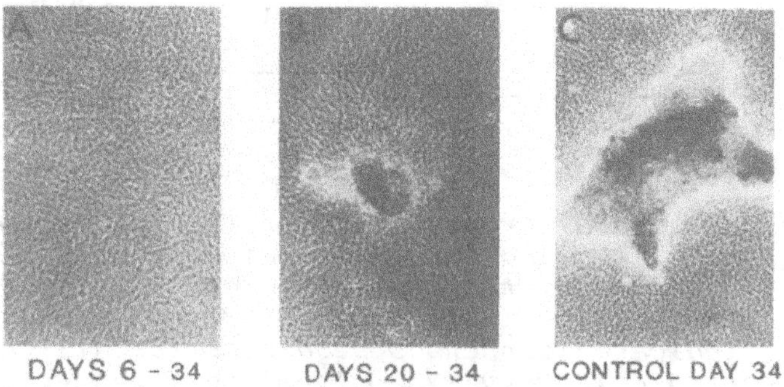

Figure 8. Phase contrast micrographs (200x) of rat osteoblast cultures 34 days after plating. *(A) Cultures treated continuously with 10^{-8}M $1,25(OH)_2D_3$ (every media change) from Day 6 appear to have limited multi-layering of cells and inhibition of nodule formation as compared to the control (C). (B) Cultures were treated continuously with hormone from Day 20, after development of mineralized nodules. Note the smaller size nodules compared to the nodule size in the control culture (C) at Day 34 suggesting inhibited growth of the nodule from Days 20-34 in the presence of hormone.*

OSTEOCALCIN, A GENE REGULATED AS A FUNCTION OF CELL GROWTH AND DIFFERENTIATION

General Features of the Gene

The osteocalcin gene (Figure 9) encodes a 10 K Da polypeptide, the osteocalcin precursor which is processed intracellularly to a 6,000 molecular weight protein that represents one of the most abundant non-collagenous proteins accumulated in bone tissue (38-39). It is a protein characterized by the presence of the vitamin K dependent amino

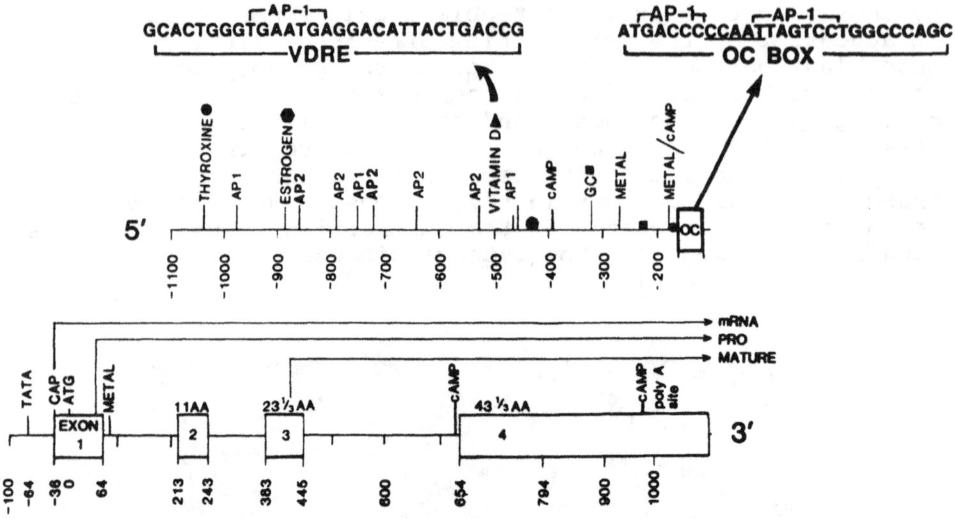

Figure 9. ***Structural organization of the rat osteocalcin gene, showing intron and exon organization and location of the propeptide and mature osteocalcin.*** *In the 1100 nucleotide, 5' flanking sequences are indicated consensus sequences for all genes transcribed by RNA polymerase II, several steroids known to alter osteocalcin synthetic levels and other modifiers (cAMP). Sequences of two regulatory elements that have been defined and partially characterized, the vitamin D responsive element (VDRE) and the osteocalcin (OC) box, the primary transcription regulatory element containing the CCAAT element as a central motif are shown. Both these regulatory elements were identified by DNAse footprinting, gel retardation assays and DMS interference analysis, as described in Markose et al., 1990. Within these two elements are also found active AP-1 sites which bind to the oncogene encoded fos-jun proteins. Since osteocalcin is not expressed in proliferating osteoblasts and cannot be induced by vitamin D, we propose these AP-1 sites serve as a regulatory mechanism to keep the gene repressed during the proliferative period until other signals activate gene transcription postproliferatively (Owen et al., Proc. Natl. Acad. Sci. USA, in press; Lian et al., J. Cell. Biochem., in press).*

acid γ-carboxyglutamic acid necessary for its calcium binding properties (39). To date it is the only known bone specific protein and appears to function in regulation of the mineral phase of bone (40). Both the human (41-42) and rat osteocalcin genes have been cloned (26-28) and characterized and have similar general organization. The human osteocalcin gene has been localized to the q region of chromosome 1 (43) (Figure 10). Also located h. chromosome 1 is the bone phenotype alkaline phosphatase gene (44).

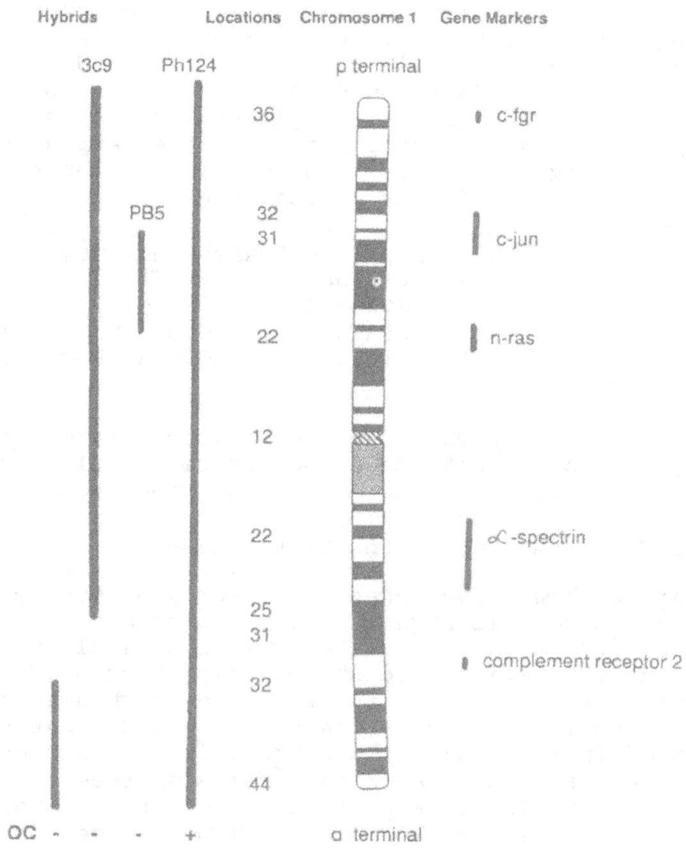

Figure 10. Regional localization of the osteocalcin gene on human chromosome 1. *The region on chromosome 1 where the osteocalcin gene is located was determined by Southern blot analyses of DNA from mouse-human somatic cell hybrids (3C9, PB5, Ph124) that retain segments of chromosome 1. These have been identified by use of cloned probes which were previously assigned to specific regions of chromosome 1. The OC gene resides at the q terminal end (Puchacz et al. Endocrinology 124: 2648-2650, 1989).*

The rat osteocalcin gene promoter, regulated by vitamin D, has been studied in great detail (26-29,41-43,45-47). A modular organization of the gene is evident (26) from the presence of (1) consensus sequences (SP1, AP2, TATA) for all genes transcribed by RNA polymerase II; (2) primary transcriptional elements including the OC Box, a highly conserved region of 24 nucleotides between the human and rat genes that contains the CAAT motif as the central element and flanked by adjacent AP-1 sites; (3) several AP-1 sites that bind the jun-fos protein complex encoded by the oncogenes c fos and c jun, and (4) a series of promoter elements responsive to steroid hormones, including glucocorticoid, estrogen, Vitamin D and the nucleotide cAMP which are all physiologic regulators of osteocalcin synthesis. While mRNA levels for osteocalcin and transfection of a chimeric gene construct (OC promotor-chloramphenicol acetyl transferase (CAT) gene) into osteoblasts show CAT enzyme activity reflecting transcriptional regulation of the gene by glucocorticoid (29), cAMP (48) and vitamin D (26-29,42,45-46). To date, only the vitamin D responsive element has been identified at single nucleotide resolution in the OC gene (42,45-46). The steroid-hormone responsive elements which may be active in the OC gene promoter all fall into a closely related family of sequences, as shown below.

Hormone	Consensus Sequences		
$1,25(OH)_2D_3$	CTGGGTGAATGA	GGACA	TTACTGA
Estrogen	AGGGTCA	C A G	TGACCT
Thyroxine	AGATCAA	GGG ACG	TGACCG
Glucocorticoid Progesterone	TGTACA	G G A	TGTTCT

Recent studies suggest multiple steroid recognition of the element, for example, between progesterone and glucocorticoid (49) and may explain opposing effects on gene regulation.

Protein (Nuclear Factors)--DNA Interactions

One approach we have taken to directly study the regulation of the osteocalcin gene at various stages throughout the developmental sequence was to identify the elements involved in the regulation through determinations of protein-DNA interaction in the 5' regulatory elements which determine the specificity and level of gene transcription. Nuclear extracts were prepared from normal diploid proliferating osteoblasts that do not express osteocalcin and cells expressing high levels of osteocalcin isolated from the mineralized matrix (Figure 11A). These extracts were examined in gel mobility shift assays with gene segments of the osteocalcin gene promoter. As shown in Figure 11B, within the first 600 nucleotides of the parameter, different protein-DNA interactions are observed in the two extracts. Using a smaller segment of the OC gene promoter (approximately 200 nucleotides, Figure 11C) that spanned nucleotides -527- 344, which includes the VDRE and AP-1 sites, clearer specific interactions are seen. The presence and absence of bands (protein-DNA complexes) in nuclear extracts from proliferating osteoblasts compared to mineralizing cultures demonstrated specific interactions reflecting both negative and positive regulation of the gene. For example, in the proliferating osteoblasts

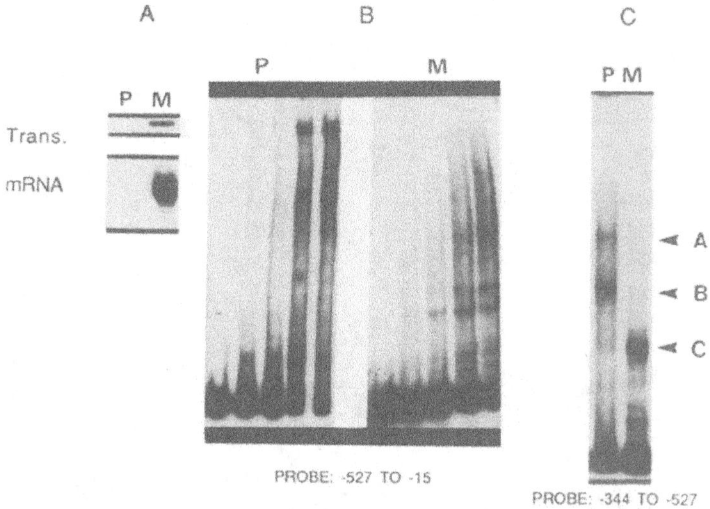

Figure 11. Protein-DNA interactions in the osteocalcin promoter in nuclear extracts from proliferating (P) normal diploid rat osteoblasts and from cells harvested at day 28 in a mineralized matrix (M). *Panel A shows the presence of osteocalcin transcription (T) and mRNA accumulation in differentiated osteoblasts and absence of expression in proliferating osteoblasts (Day 8). In Panel B, a large gene segment (-15 to -527), which is not usually used in gel retardation assays, was used as a probe; however, protein-DNA complexes (bands) are evident, reflecting interactions in proliferating cells that are absent in cells from mineralizing cultures, denoted by the upper brackets, as well as interactions present only in differentiated cells (lower bracket). In Panel C, using a smaller segment of the probe that includes the VDRE and several AP-1 sites, clearer interactions are evident with bands reflecting protein-DNA complexes found in nuclear extracts from proliferating cells (bands A and B, negative regulatory elements) that are absent from cells in mineralized cultures when the gene is transcribed. Band C clearly reflects a positive regulatory element as it is absent from nuclear extracts from the proliferating cells.*

which do not express osteocalcin, bands are present that are not found in nuclear protein extracts from cells in mineralizing cultures (negative regulation). In extracts from mineralizing cultures, bands are present when the gene is transcribed that are not found in proliferating extracts, indicating positive regulatory elements.

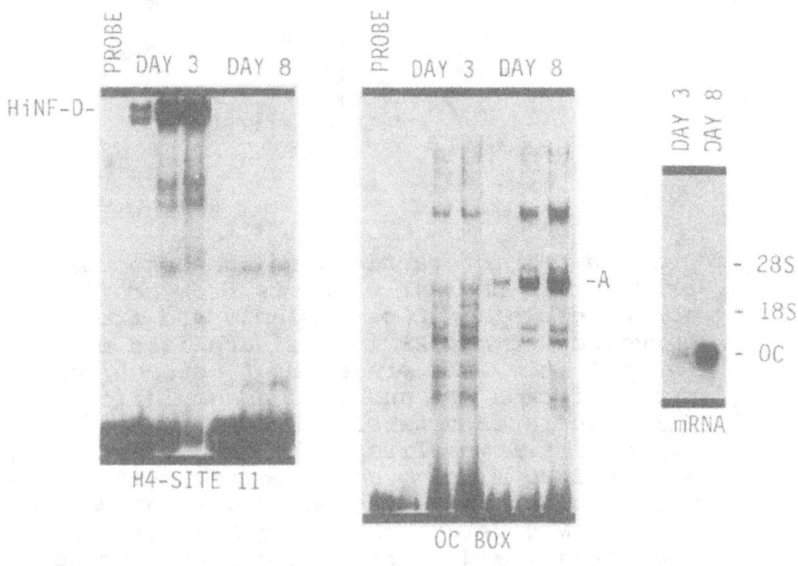

Figure 12. Reciprocal relationship between the binding of nuclear factors to promoters of a proliferation-coupled H4 histone gene and the differentiation-associated osteocalcin gene. *ROS 17/2.8 osteosarcoma cells were harvested while proliferating (day 3) or following confluence (day 8) and nuclear protein extracts and total cellular RNA prepared. (A) Binding of nuclear factor HiNF-D to the Site II region of the F0108 H4 histone gene promoter as determined by gel retardation assay, greatly decreased on day 8. (B) Binding of nuclear proteins to the osteocalcin gene promoter OC box, determined by gel retardation assay, shows an increased protein-DNA interaction (band A, characterized in reference 7) upon confluence at day 8. For panels A and B, the three lanes for each day represent 2.5, 5, or 7.5 µg nuclear protein per lane. (C) Osteocalcin mRNA (OC), as determined by Northern blot analysis of total cellular RNA, shows a 10-fold increase between day 3 and day 8, consistent with transcriptional activation (Panel B). 28S and 18S indicate the positions of the 28s and 18s ribosomal RNAs.*

The reciprocal relationship between expression of the phenotypic genes and the proliferative state of the cell is somewhat deregulated in osteosarcoma cells (7,50 and previous chapter) and OC is expressed in such proliferating cells (Figure 12) in contrast to normal diploid rat osteoblasts (Figure 11A). However, even in tumor cell cultures when confluency is reached, histone gene transcription ceases and a concomitant ten fold increase in osteocalcin mRNA and synthesis occurs with an increase in protein-DNA interactions at the OC box. This suggests normal transcriptional regulation of histone and OC gene expression, even in the tumor cells, but only at confluency, when it is nondividing.

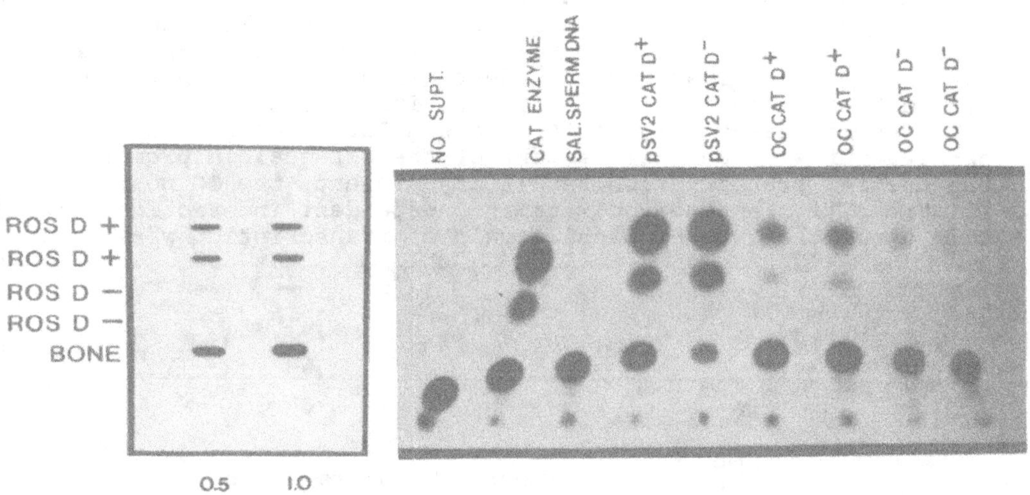

Figure 13. Example of deletion analysis of the osteocalcin promoter to locate vitamin D responsiveness using ROS 17/2.8 cells. (Left) Osteocalcin mRNA levels in ROS 17.2 cells pretreated with 10 nM 1,25(OH)$_2$D$_3$ (D+) and controls (D−). *Total cellular RNA (0.5 or 1.0 µg) was subjected to slot blot analysis using a ^{32}P-labeled osteocalcin gene hybridization probe.* ***(Right) CAT assay using cell extracts from ROS 17.2 cells, transfected with chimeric DNA consisting of 600 nucleotides of the rat osteocalcin 5' proximal regulatory sequences fused to the CAT structural gene.*** *Arrow designates acetylated forms of chloramphenicol resulting from CAT activity. Lanes: No supt, negative control for the CAT assay (no cell extract in the reaction mixture); CAT enzyme (positive control); Cells transfected with salmon sperm DNA (control); pSV2 CAT D$^+$, CAT activity in extracts from 1,25(OH)$_2$D$_3$-treated cells transfected with pSV2 CAT; pSV2 CAT D$^-$, same as previous lane but without addition of 1,25(OH)$_2$D$_3$; OC CAT D$^+$, CAT activity in extract from 1,25(OH)$_2$D$_3$-treated cells transfected with osteocalcin-CAT chimeric DNA; OC CAT D$^-$, same as lanes OC CAT D$^+$ but in the absence of 1,25(OH)$_2$D$_3$.*

Very distinct protein-DNA interactions occur in the osteocalcin promoter with extracts from cells treated with vitamin D and such bands were characterized to identify the vitamin D responsive element. Initially we selected a region of the OC promotor within the first 600 nucleotides (based on data from deletion constructs and CAT assays, Figure 13) that contained steroid hormone-like sequences (that is, bearing homology to other known steroid responsive elements glucocorticoid, estrogen and thyroxine) and was responsive to vitamin D.

Specific interactions were first identified by DNase I footprint analyses (Figure 14). A region from -437 to -465 clearly showed specific binding of nuclear proteins from vitamin D treated bone cells that resulted in accessibility of DNase to cleave nucleotides in that region. Vitamin D receptor complex binding to this region was further characterized by protein-DNA interactions in the gel retardation assays and by specific competition with an oligonucleotide of the 24 nucleotide putative VDRE segment (Figures 15A and B). The site-specific G nucleotide-protein contacts of the vitamin D dependent bands observed in the gel retardation assays were further identified by the DMS interference assay (Figure 16). A similar strategy was carried out with another segment of the osteocalcin promoter, the primary transcription regulatory element, the OC box (Figure 17). We observed vitamin D dependent interactions at this element only when basal levels of transcription were low.

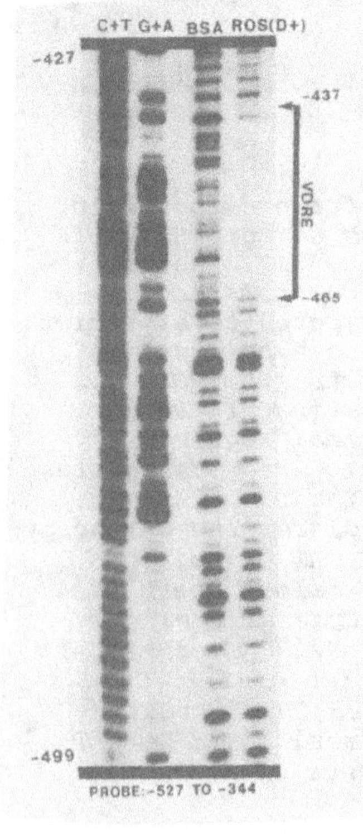

*Figure 14. **DNase I footprint analysis of protein-DNA interactions in the -527 to -344 promoter region of the rat osteocalcin gene.** C + T (lane 1) and G + A (lane 2) sequencing reactions representing the coding strand were electrophoresed along with the DNase I footprint reactions. The control lane (lane 3) shows the DNase I digestion pattern of the probe incubated with 3 μg of bovine serum albumin. The pattern of digestion obtained with 13 μg of nuclear extract from 1,25(OH)$_2$D$_3$-treated ROS 17/2.8 cells (lane 4) shows a specific region of protection indicated by solid line and designated VDRE.*

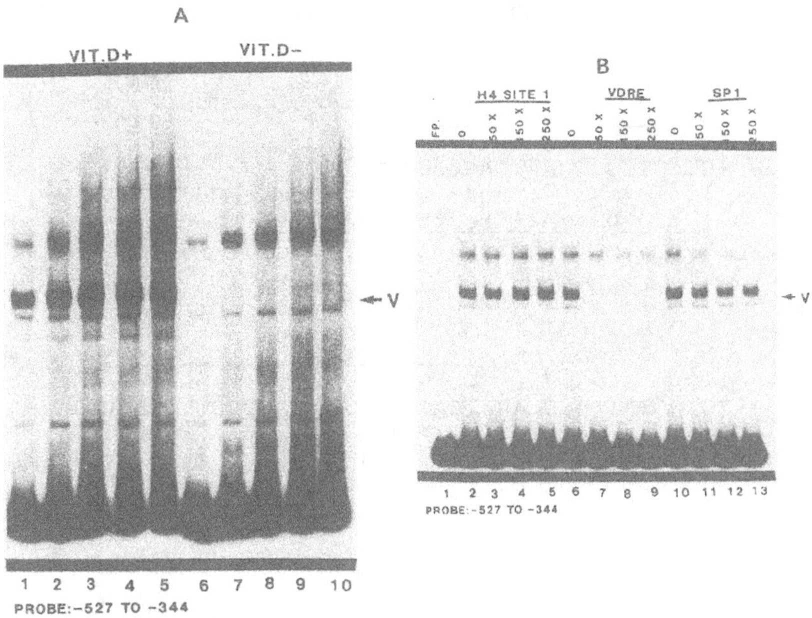

Figure 15. ***(A) Detection of factors with affinity for the osteocalcin promoter region -527 to -344 by gel mobility-shift analysis.*** *Increasing amounts of nuclear extracts (2.5, 5, 7.5, 10 and 12 μg) of 1,25(OH)$_2$D$_3$ -treated (VIT.D+; lanes 1-5) and untreated (VIT>D-; lanes 6-10) ROS 17/2.8 cells were used in standard DNA binding reactions. The major protein-DNA complex that appears with vitamin D$_3$ treatment is denoted V.* ***(B) Competition analysis of DNA-protein complexes in the HincII/Bgl II (-527 to -344) fragment of the rat osteocalcin promoter.*** *Five micrograms of nuclear extract from vitamin D-treated ROS 17/2.8 cells was added to standard binding reaction mixtures. Reaction mixtures also contained either no additional DNA (lanes 2, 6 and 10) or 50-fold (lanes 3, 7, and 11), 150-fold (lanes 4, 8 and 12), and 250-fold (lanes 5, 9, and 13) of an unlabeled synthetic oligonucleotide. The oligonucleotide used as the competitor in lanes 3-5, designated as H4 site I, contains a cAMP binding site. A partial homology to this consensus sequence is seen at -453 to -446 of the osteocalcin promoter. The oligonucleotide used as the competitor in lanes 7-9, designated VDRE, corresponds to -462 to -436 of the rat osteocalcin promoter. In lanes 11-13, the competitor used was the oligonucleotide synthesized to correspond to high-affinity binding sites for Sp1.*

The vitamin D dependent nuclear protein factor-DNA interactions at the VDRE and OC box, defined by the DNase I and DMS analyses, are summarized below:

VDRE Probe

```
-517                    VDRE                     -420
              _______________________________
              |                             |
5'----------GCACTGGGTGAATGAGGACATTACTGAGC--------3'
            |   •••        ••       •      |
            |_______DNase 1 Footprint______|
```

OC Box

```
-123                              OC Box                          -65
                         ___________________________________
                         |                                 |
5'---------TGACCTATTGCGCACATGACCCCCAATTAGTCCTGGCAGCAT-------3'
           |                   •••▲▲                |
           |_____________DNase 1 Footprint__________|
```

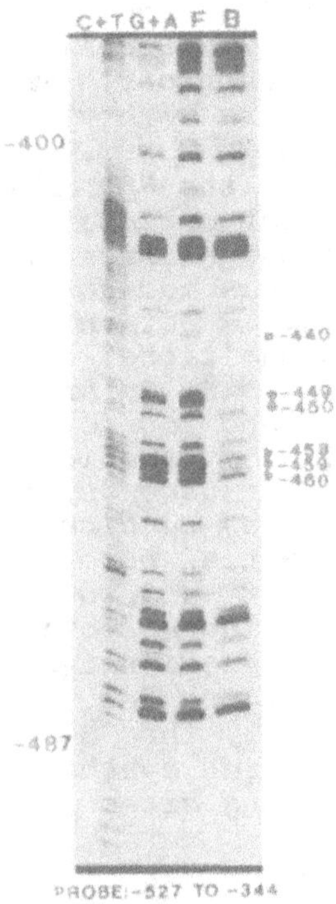

Figure 16. Methylation interference analysis was performed to establish the contact sites of the vitamin D-induced factor. *The ^{32}P-end-labeled probe (-527 to -344; pOC3.4) was partially methylated with dimethylsulfate and used in standard DNA binding reactions. Nuclear extracts from vitamin D-treated ROS 17/2.8 cells were added and the DNA-protein complexes were resolved by native gel electrophoresis. DNA from the protein-DNA complex, indicated as V in A, and from free probe were cluted and cleaved with piperidine. An equal number of counts of free (lane 3) and bound (lane 4) DNA were electrophoresed on 8% polyacrylamide denaturing gels. Sequencing reactions C + T (lane 1) and G + A (lane 2) of the coding strand were electrophoresed alongside. The G residues that interfere with the binding of the protein are indicated by solid circles.*

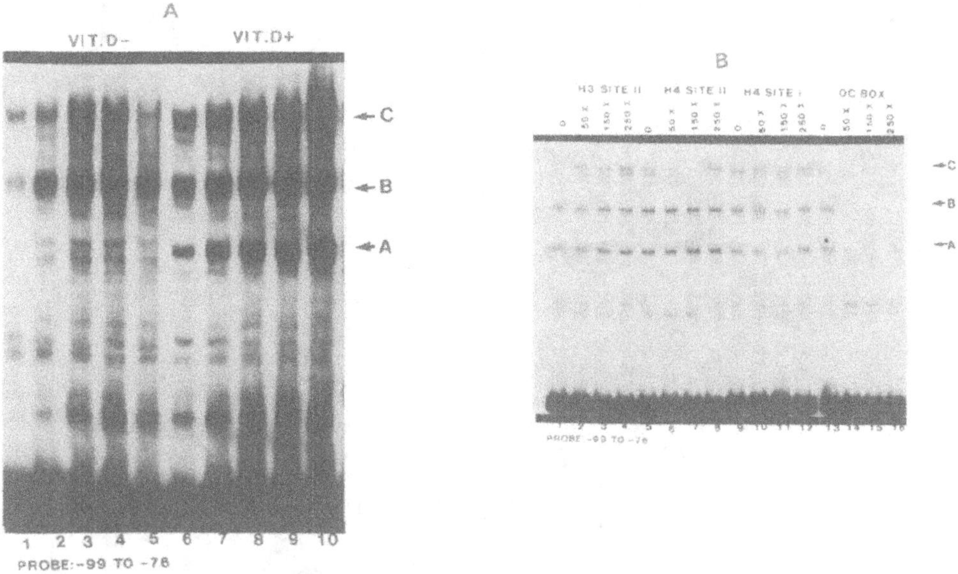

Figure 17. Vitamin D Dependent Protein-DNA Interactions at the OC Box Promoter Element. *(A) Nuclear extracts from $1,25(OH)_2D_3$-treated (VIT.D+) and control (VIT.D-) ROS 17/2.8 cells were analyzed by gel mobility-shift assay, for factors that bound to the osteocalcin box region (-99 to -76) of the rat osteocalcin gene promoter. The probe was incubated with 2.5, 5, 7.5, 10 and 12 μg of vitamin D-lacking (lanes 1-5) and vitamin D-containing (lanes 6-10) ROS cell extracts. The major DNA-protein complexes A, B, And C are indicated. (B) Competition gel mobility-shift analysis of the specificity of protein-DNA complexes A, B, and C in the osteocalcin (OC) box. The single end-labeled HINdIII/EcoRI fragment from pUC19, which contained the 24-base-pair OC box, was used as the probe. The same unlabeled fragment (50-fold, 150-fold, and 250-fold excess) was used as the OC box competitor (lanes 13-16). Other competitors used were H4 site II oligonucleotide (lanes 5-8) and H3 site II oligonucleotide (lanes 1-4), which contain the CAAT box regions of the human histone H4 and H3 genes and H4 site I oligonucleotide (lanes 9-12), which contains a putative cAMP binding site.*

Suppression of Osteocalcin Expression in Proliferating Cells

Both the OC box and the VDRE contain overlapping AP-1 sites (Fig 9). The nuclear proto-oncogene encoded proteins Jun and Fos form a stable hetero-dimeric AP-1 binding activity by leucine zipper interaction (11). It has been established that the AP-1 consensus sequences within the VDRE support sequence-specific binding of the Fos and Jun heterodimer (51,12). Our studies also show that the AP-1 consensus sequence in the OC box also actively binds purified recombinant fos-jun proteins as a complex (Figure 18). Site-specific binding is further supported by competition with a segment of human MT II_A gene promoter which contains an AP-1 binding site. Sequence-specific interactions with the AP-1

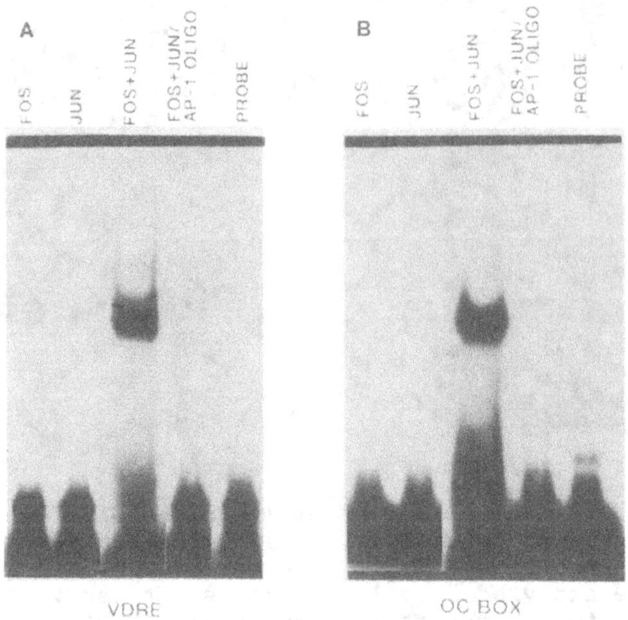

Figure 18. Binding of Purified Recombinant Fos and Jun Proteins to Rat Osteocalcin VDRE (A), OC Box (B) Probes as Assessed by Gel Retardation Assay. *For each probe, 1μM protein of either Fos alone (FOS), Jun alone (JUN), or Fos-Jun together (FOS+JUN) were incubated at 37°C for 15 minutes and 4 ng ^{32}P probe was added. Binding was allowed to occur for 15 minutes at room temperature. Protein-DNA complexes were resolved on 4.5% native polyacrylamide gels. For each probe, the binding of the Fos-Jun complex was specifically competed with an oligonucleotide spanning the human MT II_A gene AP-1 site (FOS+JUN/AP-1 OLIGO). PROBE indicates probe incubated without protein. The precise protein DNA G residue contacts identified by methylation interference analysis are designated by · in the lower panel.*

motifs were demonstrated by DMS interference patterns which define protein-DNA contacts at the guanine residues corresponding to the AP-1 consensus (12). During the normal developmental sequence of osteoblast differentiation cfos and cjun are synthesized during the proliferation period (4) and AP-1 binding activity is found at this time when osteocalcin is not expressed and is not induced by vitamin D (12). Following the downregulation of proliferation with the development of mineralized extracellular matrix, osteocalcin expression and synthesis become induced. The identification of these AP-1 sites provides a mechanism (Figure 19) to explain regulated expression and vitamin D modulation of osteocalcin only in mature non-dividing osteocytes. Thus, the overlapping of two classes of regulatory elements related to expression as a function of proliferation and expression related to basal levels of transcription and modulation by hormone activity is an example of coordinate occupancy of elements to regulate gene transcription. In the case of the osteocalcin gene, it is apparent that the AP-1 site functions as a negative regulatory element and we term this mechanism by which the OC gene is suppressed during proliferation by activation or binding of jun-fos complex to the AP-1 sites in the OC box and the VDRE as *phenotype suppression* (Figure 1C).

The possibility, therefore, can be considered that the Fos-Jun proteins which are expressed in proliferating osteoblasts contribute to regulation of other osteoblast genes. Alkaline phosphatase, for example, is also not expressed and regulated by vitamin D in proliferating osteoblasts (Figures 1 and 5) and the organization of its vitamin D VDRE sequence (52) is identical to osteocalcin in that it contains an active AP-1 site. In contrast, collagen is expressed during proliferation and is regulated by vitamin D and here the analogous VDRE sequence in the rat type αI collagen gene promotor (taken from 53) is associated with an AP-1 consensus sequence (shown by underline below) is contiguous to and does not reside within the VDRE as shown below and compared to the AP-1 site of the VDRE's for osteocalcin and alkaline phosphatase and OC box. These AP-1 sequences are compared to the well studied AP-1 site in the human metallothionein II_A promoter.

Human Metallothionein II_A AP-1 Binding Site $^{-102}$GTGACTCAGCGCG^{-90}

Rat Osteocalcin Box $^{-99}$ ATGACCCCCAATTAGTCCTGGCAGC^{-75}

Rat Osteocalcin VDRE $^{-462}$CTGGGTGAATGAGGACATTACTGA^{-439}

Human Osteocalcin VDRE $^{-510}$GGTGACTCACCGGGTGAACGGGGGCATT^{-483}

Alkaline Phosphatase VDRE $^{-948}$GGGGGTGACTGATGGTAACCTGATTG^{-923}

Type αI Collagen VDRE $^{-2957}$CTGGGGGCAGAAGAACTTTCTGGAGGATTTGAGTGA^{-2912}

Thus, it appears that subtle variations in the organization of the VDRE and AP-1 motifs in the osteoblast gene promoters may contribute to their differential expression during osteoblast phenotype development.

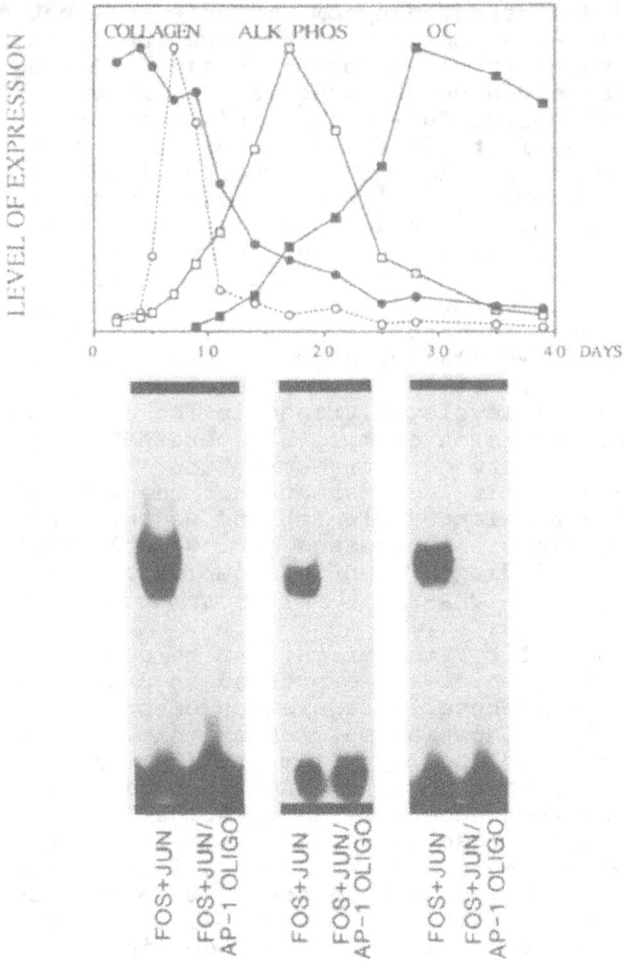

Figure 19. ***Fos-jun binding to AP-1 sites associated with vitamin-D responsive elements*** *of 3 genes expressed sequentially during osteoblast differentiation, collagen (-•-), alkaline phosphatase (-□- ALK PHOS), osteocalcin (-■- OC) (top profiles). The proliferation period is indicated by histone gene expression (---○---). Lower Panel. The sequence specific protein-DNA interactions of the oncogene encoded fos and jun protein complex and the VDRE sequences (used as probe) of each specific gene. Fos-Jun binds to AP-1 sites residing within the VDRE's of OC and AP, but to an AP-1 site contiguous to collagen VDRE. Sequence specific interactions are confirmed by competition with an oligonucleotide encoding an AP-1 sequence from the h. metallothionine gene protein. We propose this organization of AP-1 sites and VDRE's allows for Vitamin D regulation of collagen during proliferation but blocks hormone regulation of ALK PHOS and OC until the post-proliferative period.*

CONCLUSIONS

We have described an experimental model system to study the relationship between cell growth and differentiation using the developmental sequence of the osteoblast phenotype. The temporal sequence of genes characterizing different stages of osteoblast differentiation in relation to the development and organization of a bone tissue-like ECM and the associated dramatic changes in gene expression provides an opportunity to examine both transcriptional and post-transcriptional control of individual genes and signalling mechanisms that are transduced to the nucleus for the activation of specific genes at the different developmental stages. Modifications in growth and formation of the extracellular matrix demonstrate the functional relationship between development of the bone ECM and expression of osteoblast phenotypic genes. Future directions should experimentally approach more precisely mechanisms by which signals from the ECM are processed intracellularly for activation of genes. Hormonal control of bone growth and differentiation is operative and stringently regulated throughout the osteoblast developmental sequence and the effects exerted are highly dependent on the differentiated state of the cell. Likewise, the effects of factors (cytokines, growth factors, non-steroidal hormones) modulating expression of genes associated with tissue development and organization should be considered within the context of the developmental stage of the cell and tissue.

REFERENCES

1. Bissell, M.J., Hall, H.G., and Parry, G, How does the ECM direct gene expression?, J. Theoret. Biol. 99:31-68 (1981).
2. Mecham, R.P., Biology of connecticue tissue: Regulation of mesenchymal cell differentiation by ECM, in "Bone and Mineral Research," W. A. Peck, ed., 185-201 (1987).
3. Aronow, M.A., Gerstenfeld, L.C., Owen, T.A. Tassinari, M.S., Stein, G.S., and Lian, J.B., Factors that promote progressive development of the osteoblast phenotype in cultured fetal rat calvaria cells, J. Cell. Physiol. 143:213-221 (1989).
4. Owen, T.A., Aronow, M., Shalhoub, V., Barone, L.M., Wilming, L., Tassinari, M., Kennedy, M.B., Pockwinse, S., Lian, J.B., and Stein, G.S., Progressive development of the rat osteoblast phenotype in vitro: Reciprocal relationships in expression of genes associated with osteoblast proliferation and differentiation during formation of the bone extracellular matrix, J. Cell. Physiol. 143:420-430 (1990).
5. Bellows, C.G., Aubin, J.E., Heersche, J.N.M., and Antosa, M.E., Mineralized bone nodules formed in vitro from enzymatically released rat calvaria populations, Calcif. Tiss. Int. 38:143-154 (1986).
6. Bhargava, U., Bar-Lev, M., Bellows, C.G., and Aubin, J.E., Ultrastructural analysis of bone nodules formed in vitro by isolated fetal rat calvarial cells, Bone 9:155-163 (1988).

7. Holthuis, J., Owen, T.A., van Wignen, A.J., Wright, K.L., Ramsey-Ewing, A., Kennedy, M., Carter, R., Cosenza, S.C., Soprano, K.J., Lian, J.B., Stein, J.L., and Stein, G.S., Tumour cells exhibit deregulation of the cell cycle controlled, proliferation specific histone gene promoter factor HiNF-D, Science 247:1454-1457 (1990).
8. Glowacki, J. and Lian J., Impaired recruitment of osteoclast progenitors in deficient osteocalcin-depleted bone implants, Cell Differentiation 21:247-254 (1987).
9. Glowacki, J., Rey, C., Cox, K.A., and Lian, J.B., Effects of bone matrix components on osteoclast differentiation, Conn. Tiss. Res. 21:121-129 (1989).
10. Reinholt, F.P., Hultenby, K., Oldberg, Å., and Heinegård, D., Osteopontin--a possible anchor of osteoclasts to bone, Proc. Natl. Acad. Sci. USA 87:4473-4475 (1990).
11. Rauscher III, F.J., Sambucetti, L.C., Curran, T., Distel, R.J., and Spiegelman, B.M., Common DNA binding site for Fos protein complexes and transcription factor AP-1, Cell 52:471-478 (1988).
12. Owen, T.A., Bortell, R., Yocum, S.A., Smock, S.L., Zhang, M., Abate, C., Shalhoub, V., Aronin, N., Wright, K.L., van Wijnen, A.J., Stein, J.L., Curran, T., Lian, J.B., and Stein, G.S. Coordinate Occupancy of AP-1 Sites in the Vitamin D Responsive and CCAAT Box Elements by Fos-Jun in the Osteocalcin Gene: A Model for Phenotype Suppression of Transcription, Proc. Natl. Acad. Sci. USA, in press.
13. Glimcher, M.J., Mechanism of calcification: Role of collagen fibrils and collagen-phosphoprotein complexes in vitro and in vivo, Anat. Rec. 224:139-153 (1989).
14. Gerstenfeld, L.C., Chipman, S.D., Glowacki, J., and Lian, J., Expression of differentiated function by mineralizing cultures of chicken osteoblasts, Dev. Bio. 122:49-60 (1987).
15. Lian, J.B., Roufosse, A.H., Reit, B., and Glimcher, M.J., Concentrations of osteocalcin and phosphoprotein as a function of mineral content and age in cortical bone, Calcif. Tiss. Int. 34:S82-S87 (1982).
16. Gerstenfeld, L.C., Chipman, S.D., Kelly, C.M., Hodgens, K.J., Lee, D.D., and Landis, W.J., Collagen expression, ultrastructural assembly, and mineralization in cultures of chicken embryo osteoblasts, J. Cell. Biol. 106:979-989 (1988).
17. Jones, S.J. and Boyde, A., Morphological changes of osteoblasts in vitro, Cell Tissue Res. 166:101-107 (1976).
18. Holtrop, M.E., The Osteoblast and Osteocyte in "Bone, A Treatise," B.K. Hall, ed., (1991).
19. Bruder, S.P. and Caplan, A.I., Terminal differentiation of osteogenic cells in the embryonic chick tibia is revealed by a monoclonal antibody against osteocytes, Bone 11:189-198 (1990).
20. Weinreb, M., Shinar, D., and Rodan, G., Different pattern of alkaline phosphatase, osteopontin, and osteocalcin expression in developing rat bone visualized by in situ hybridization, J. Bone Min. Res. 8:831-842 (1990).
21. DeLuca, H.F., The vitamin D story: A collaborative effort of basic science and clinical medicine, FASEB J. 2:224-36 (1987).
22. Minghetti, P.P. and Norman, A.W., 1,25$(OH)_2D_3$-vitamin D_3

receptors: gene regulation and genetic circuitry, FASEB J. 2:3043-3053 (1989).

23. Harrison, J.R., Petersen, D.N., Lichtler, A. C., Mador, A.T., Rowe, D.W., and Kream, B.E., 1,25-Dihydroxyvitamin D_3 inhibits transcription of Type I collagen genes in the rat osteosarcoma cell line ROS 17/2.8, Endocrinology 125:327-333 (1989).
24. Manolagas, S.C., Burton, D.W., and Deftos, L.J., 1,25-Dihydroxyvitamin D_3 stimulates the alkaline phosphatase activity of osteoblast-like cells, J. Biol. Chem. 256:7115-7117 (1981).
25. Prince, C.W. and Butler, W.T., 1,25-Dihydroxyvitamin D_3 regulates the biosynthesis of osteopontin, a bone-derived cell attachment protein, in clonal osteoblast-like osteosarcoma cells, Collagen Rel. Res. 7:305-313 (1987).
26. Lian, J., Stewart, C., Puchacz, E., Mackowiak, S., Shalhoub, V., Collart, D., Zambetti, G., and Stein, G., Structure of the rat osteocalcin gene and regulation of vitamin D-dependent expression, Proc. Natl. Acad. Sci. USA 84:1143-1147 (1989).
27. Yoon, K., Rutledge, S.J., Buenaga, R.F., and Rodan, G.A., Characterization of the rat osteocalcin gene: Stimulation of promoter activity by 1,25-dihydroxyvitamin D_3, Biochemistry 27:8581- 8526 (1988).
28. Demay, M.B., Roth, D.A. and Kronenberg, H.M., Regions of the rat osteocalcin gene which mediate the first of 1,25-dihydroxyvitamin D_3 on gene transcription, J. Biol. Chem. 264:2279-2282 (1989).
29. Morrison, N.A., Shine, J., Fragonas, J.-C., Verkest, V., McMenemy, L., and Eisman, J.A., 1,25-dihydroxyvitamin D-responsive element and glucocorticoid respression in the osteocalcin gene, Science 246:1158-1161 (1989).
30. Fraser, J.D., Otawara, Y., and Price, P.A., 1,25-Dihydroxyvitamin D_3 stimulates the synthesis of matrix γ-carboxyglutamic acid protein by osteosarcoma cells, J. Biol. Chem. 263:911-916 (1988).
31. Lian, J.B., Carnes, D.L., and Glimcher, M.J., Bone and serum concentrations of osteocalcin as a function of 1,25-dihydroxyvitamin D_3 circulating levels in bone disorders in rats, Endocrinology 120:2123-30 (1987).
32. Suda, T., Shinki, T., and Takahashi, N., The role of vitamin D in bone and intestinal cell differentiation, Annu. Rev. Nutr. 10:195-211 (1990).
33. Majeska, R.J. and Rodan, G.A., The effects of $1,25(OH)_2D_3$ on alkaline phosphatase in osteoblastic osteosarcoma cells, J. Biol. Chem. 257:13362-13365 (1982).
34. Spiess, Y.H., Price, P.A., Deftos, J.L., and Manolagas, S.C., Phenotype-associated changes in the effects of 1,25-Dihydroxyvitamin D_3 on alkaline phosphatase and bone Gla-protein of rat osteoblastic cells, Endocrinology 118:1340-1346 (1986).
35. Franceschi, R.T., Romano, P.R., and Park, K.Y., Regulation of type I collagen synthesis by 1,25-dihydroxyvitamin D_3 in human osteosarcoma cells, J. Biol. Chem. 263:18938-45 (1988).
36. Beresford, J.N., Gallagher, J.A., and Russell, R.G.G., 1,25-dihydroxyvitamin D_3 and human bone-derived cells *in vitro*: effects on alkaline phosphatase, type I collagen and proliferation, Endo. 119:1776-1785 (1986).

37. Canalis, E. and Lian, J.B., 1,25-Dihydroxyvitamin D_3 effects on collagen and DNA synthesis in periosteum and periosteum-free calvaria, Bone 6:457-460 (1985).
38. Gundberg, C.M., Hauschka, P.V., Lian, J.B., and Gallop, P.M., Osteocalcin isolation, characterization, and detection, Methods Enzymol. 107:516-544 (1984).
39. Hauschka, P.V., Lian, J.B., Cole, D.E.C., and Gundberg, C.M., Osteocalcin and matrix Gla protein: vitamin K-dependent proteins in bone, Physiologic Reviews, 990-1047 (1989).
40. Lian, J.B., Osteocalcin: Functional studies and postulated role in bone resorption, in "Current Advances in Vitamin K Research," J.W. Suttie, ed., Elsevier Science Publishers, New York, p. 245-257 (1987).
41. Celeste, A.J., Rosen, V., Buecker, J.L., Kriz, R., Wang, E.A., and Wozney, J.M., Isolation of the human gene for bone Gla protein utilizing mouse and rat cDNA clones, EMBO J. 5:1885-1890 (1986).
42. Kerner, S.A., Scott, R.A., and Pike, J.W., Sequence elements in the human osteocalcin gene confer basal activation and inducible response to hormonal vitamin D_3. Proc. Natl. Acad. Sci. USA 86:4455-4459 (1989).
43. Puchacz, E., Lian, J.B., Stein, G.S., Wozney, J., Huebner, K., and Croce, C., Chromosomal localization of the human osteocalcin gene, Endocrinology 24:2648-2650 (1989).
44. Swallow, D.M., Mapping of the gene coding for the human liver/bone/kidney isozyme of alkaline phosphatase to chromosome 1, Ann. Human Genetics 60:229-235 (1986).
45. Markose, E.R., Stein, J.L., Stein, G.S., and Lian, J.B., Vitamin D-mediated modifications in protein-DNA interactions at two promoter elements of the osteocalcin gene, Proc. Natl. Acad. Sci. USA 87:1701-1705 (1990).
46. Demay, M.B., Gerardi, J.M., DeLuca, H.F., and Kronenberg, H.M., DNA sequences in the rat osteocalcin gene that bind the 1,25-dihydroxyvitamin D_3 receptor and confer responsiveness to 1,25-dihydroxyvitamin D_3. Proc. Natl. Acad. Sci. USA 87:369-373 (1990).
47. Price, P.A. and Baukol, S.A., 1,25 Dihydroxy vitamin D_3 increases synthesis of the vitamin K dependent bone protein by osteosarcoma cells., J. Biol. Chem. 255:11660-11665 (1980).
48. Noda, M., Transcriptional regulation of osteocalcin production by transforming growth factor-β in rat osteoblast-like cells, Endocrinology 124:612-617 (1989).
49. Ankenbauer, W., Stråhle, U., and Schütz, G., Synergistic action of glucocorticoid and estradiol responsive elements, Proc. Natl. Acad. Sci. USA 85:7526-7530 (1988).
50. Wingender, E., Transcription regulating proteins and their recognition sequences, Eukaryotic Gene Expression 1:11-48 (1989).
51. Schüle, R., Kazuhiko, U., Mangelsdorf, D.J., Bolado, J., Pike, J.W., and Evans, R.M., Jun-fos and receptors for vitamins A and D recognize a common response element in the human osteocalcin gene, Cell 61:497-504 (1990).
52. Matsuura, S., Kishi, F., and Kajii, T., Characterization of a 5'-flanking region of the human liver/bone/kidney alkaline phosphatase gene: two kinds of mRNA from a single gene, Biochem. Biophys. Res. Comm., 168:993-1000 (1990).

53. Lichtler, A., Stover, M.L., Angilly, J., Kream, B., and Rowe, D.A., Isolation and characterization of the rat α1 (I) collagen promoter, J. Biol. Chem. 264:3072-3077 (1989).

This work is supported by grants from the National Institutes of Health (AR33920, AR35166, AR39588), the International Life Sciences Institute and the Northeast Osteogenesis Imperfecta Society.

THE BEGINNING OF A MOLECULAR DESCRIPTION OF HUMAN CANCER: CHROMATIN 3D-STRUCTURE, DNA SUPERCOIL, H1 HISTONE, MICROTUBULES AND THE MODULATION OF UNBALANCED GENE EXPRESSION

C. Nicolini

Institute of Biophysics Medical School University of Genova (Italy)
Salita Superiore della Noce, 35 - 16132 GENOVA

Time ago[28] several experimental evidences on the higher order chromatin structure led to the introduction of a single repeating structural unit called "fibrosome" and to its equation to a single gene.

Its detailed in situ 3-D organization pointed out the existence of fixed sites at both ends of each fibrosome (or "loop") subunit, which becomes apparent by non-invasive biophysical probes at the level of both interphase chromatin[28] and of a single metaphase chromosome depleated of all the histone proteins; this led to the suggestion of a DNA supercoil locally maintained and modulated through complex interactions with ions and chromosomal proteins[30-34-35-36], which was compatible with earlier titration studies with intercalating dyes on native isolated chromatin[24].

In recent times a consensus has been emerging on the critical role of DNA supercoil in the effective transcription of human genome in normal and cancer cells[7-52].

An alternative mechanism in the control of abnormal gene expression and cell growth was identified in the disruption of microtubules organization and orientation[8], and in the related coupling of cell to nuclear geometry[26]. The generalized repression of most genes is the key feature of human genome (highly acidic), made apparently possible by the presence of basic chromosomal proteins, namely the histone H1. Up to now there appears to be a wide consensus on the fact that the most likely candidates in the modulation of selective oncogenes expression are the removal or the enzymatic modifications of the above histone (namely its phosphorilation). Also the acetylation of H4 and the removal of the related HMG non-histones appear to play a role.

In the human cell repression is very efficient since transcription is suppressed by a factor 10^8-10^9 over the fully active state - six order of magnitude higher than in prokaryotes. In this review we intend to emphasize problems still open, paradoxes and puzzles still unsolved rather than limit ourselves to a detailed account of facts firmly established and already reviewed by us[33-34-35] and others[54-51].Within this context we present also a brief summary of recent findings in our laboratory and we'll outline the beginning of a molecular description of the events causing a normal human cell to become a cancer cell.

Molecular Basis of Human Cancer
Edited by C. Nicolini, Plenum Press, New York, 1991

CHROMATIN

Eleven years ago[25] several physico-chemical and functional changes of chromatin in a wide variety of normal transformed cells, were shown to be compatible with the existence in native chromatin of a two-order superhelical structure (Figure 1).

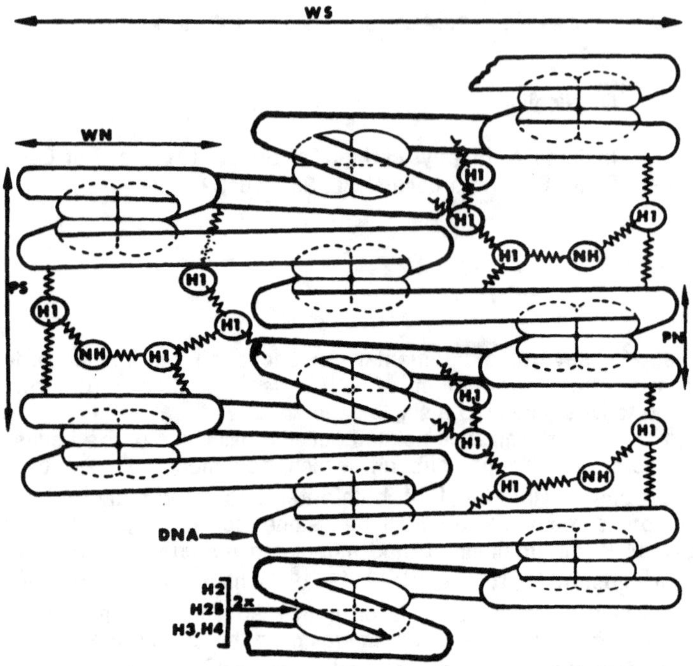

Figure 1. Two-order superhelix model for native chromatin structure[24].

This suggestion made originally in 1975 by my group to support the controversial concept of supercoiled DNA[23-24] has since then represented an alternative view to the later but more popular "solenoid" model by Klug group[14].
The differential light scattering, the melting and premelting properties, the viscosity, the number of binding sites for ethidium bromide, linearly related to the molar ellipticity at 272 nm, the template activity proportional to primary binding sites were shown to completely fit in a coherent picture the structural features suggested by two-order superhelix for native chromatin, relating them to the functional changes detected during the cell cycle, G0-G1 transition (both in vivo and in vitro), virus transformation, aging, de-differentiation and chemical carcinogenesis.
Indeed the invariance in the structural differences monitored by the above methods between any two given native chromatins (from any two cell lines of different

functional properties) before and after shearing, can be explained only by the existence in the native chromatin of a packed DNA tertiary structure in nucleosomes which is folded into "quaternary" polynucleosomal superhelix whose periodicity (t_s) is dependent upon that of the "tertiary" nucleosomal superhelix (t_n), i.e. $t_s = K\, t_n$. In such "two-order superhelix" the shearing, by breaking the DNA strands, would disrupt the quaternary but mantain intact the tertiary DNA supercoiling. Among the several predictions arising from the above model and from low EB titration studies[25], the one directly relating the DNA linker to the diameter of the corresponding chromatin fiber was strikingly confirmed even numerically by very recent cryo-electron microscopy of frozen-hydrated chromatin fibers from sea cucumber and erythrocytes[2].

For a linker length of 48 and 87 bp the predicted value was respectively 348 and 521 Å, against the 320 ± 30 and 435 ± 42 Å experimentaly determined; at the same time the original solenoid model[14] was erroneously predicting a constant fiber diameter.

Considering the crude assumptions being then made (4 nucleosomes per superhelical turn with constant 110 Å pitch and with the unit vector normal to the nucleosome face being oriented exactly parallel to the solenoidal helical axis), our early model proves still surprisingly accurate in predicting from experimentally determined negative superhelical density variable linker lenght and variable fiber width depending on tissue and species.

An updating of our model with the correct pitch, nucleosomes number (six per turn, from neutron data on mass per unit length) and orientation (its normal vector perpendicular to the "solenoidal" axis, from the presence of a CIDS signal) still mantain an excellent agreement.

In principle, each of the "variable diameter" models subsequently introduced for chromatin, like the cross-linker model characterized by nonsequential arrangements of nucleosome cores connected by transverse linker DNA[46-4-5], reconciled with local variations in linker length; including the old "solenoid" model which can be modified ad hoc to account for a variable diameter. Several other predictions however are equally contained in our early *"two-order superhelix"* model, all upheld by experimental observations a decade later, namely:

1) *the existence of fixed sites along the fiber,* made evident in multinucleofilament rope by a non-invasive EM technique on phospholipid monolayers[28], to account for the observed negative superhelicity;

2) *heterogeneity of linker length and fiber width* both local and regional, to account for differential gene activity and for the fiber periodicity and flexibility;

3) *the artifactual nature of a soluble chromatin prepared by limited nuclease* which has no bearing with the structure and function of native chromatin in situ which can be *preserved by the cold water isolation* procedure[33-34-35-36-37-38]. Only nucleofilaments longer than 200 nucleosomes display structural[4-5-28], thermodinamic[37-38] and optical[11-12] properties typical of native chromatin in situ, even at low ionic strength. Synchrotron radiation X-ray scattering profiles of native chromatin at very high concentration and low ionic strenght display a broad maximum around $s=0.005\ nm^{-1}$ which by micrococcal nuclease digestion is shown due to an interference between nucleosomes with an average distance of 230 Å[20]. In other words the chromatin fiber possesses at low ionic strenght already a helical superstructure with a diameter comparable to that of condensed chromatin at high ionic strenght held together by H1 histone, with a significant pitch reduction to 30-40 Å[4]. Model calculations on the superstructure of chromatin found that agreement between the calculed X-ray solution scattering patterns and the experimental observations can be reached with the assumpionts that uncondensed and condensed fibers have respectively a pitch of 33 and 3 nm, a helical diameter of 20 and 27 nm, and about 3.8 and 2.5 nucleosomes per turn.

Similarly differential scanning calorimetry of "cold water" chromatin reveals (Figure 2) at low ionic strenght the last three termal transitions present in intact nuclei, while the "nuclease" chromatin from the same rat liver cells displays the disappareance of the last transition at 375°[37-38].

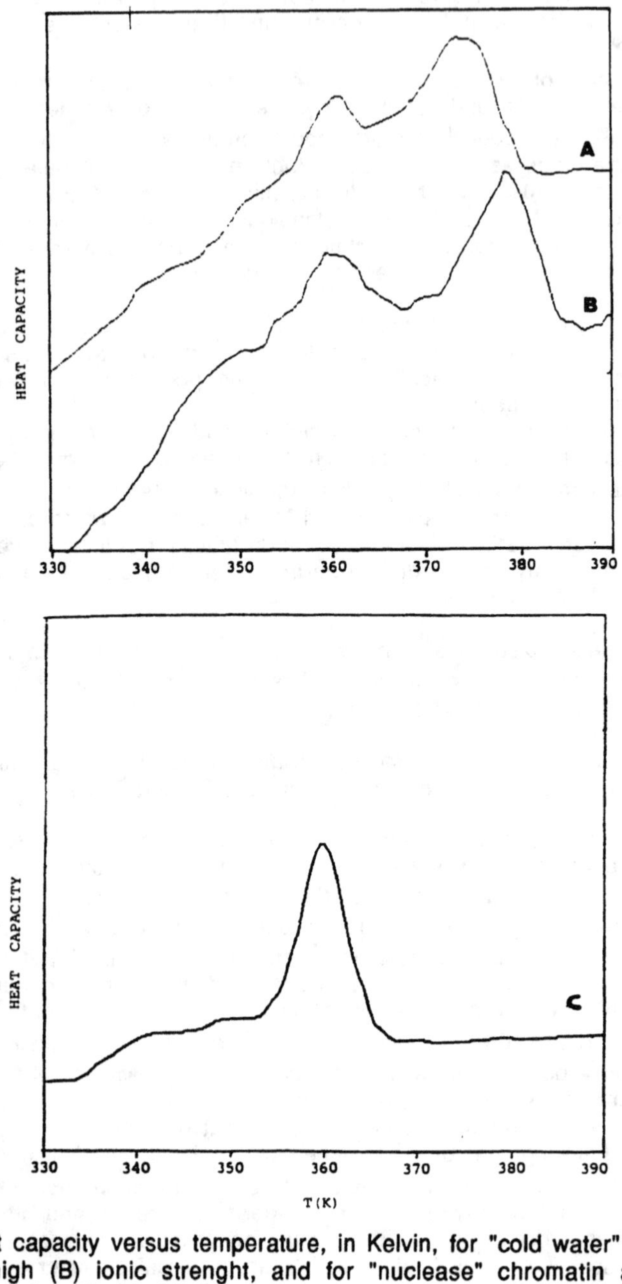

Figure 2. Heat capacity versus temperature, in Kelvin, for "cold water" chromatin of low (A) and high (B) ionic strenght, and for "nuclease" chromatin at high ionic strength (C)[33-34-35-36-37-38].

At the same time the circular intensity differential scattering[11-12] is quite pronounced for unheared "cold water" chromatin at low ionic strenght and goes nearly to zero after shearing or for the soluble chromatin isolated by limited nuclease digestion (Figure 3).

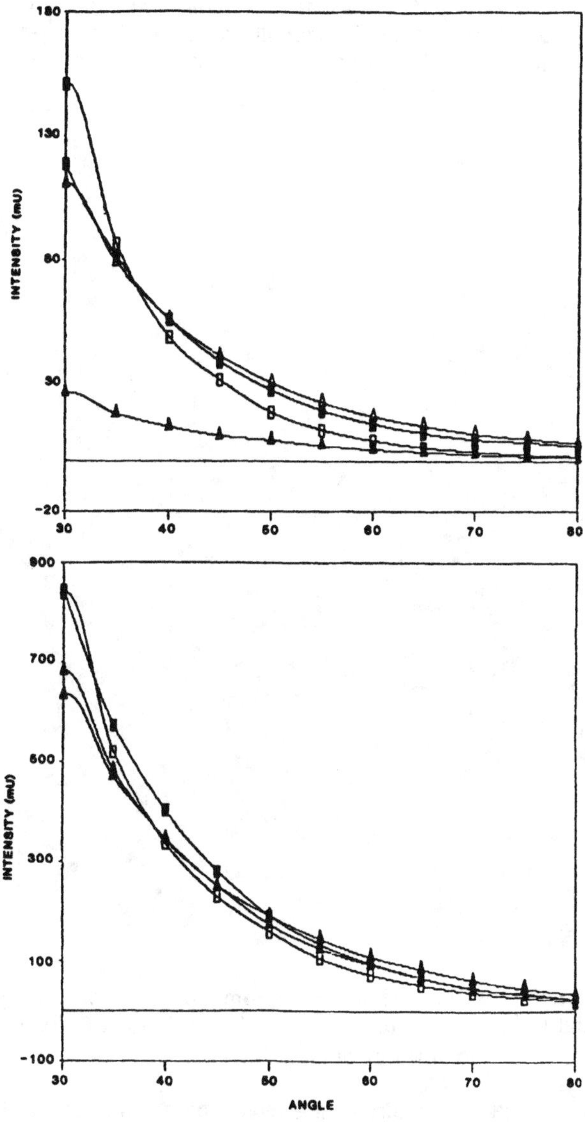

Figure 3. Above: Circular intensity differential scattering versus scattering angle of "cold water" chromatin at high (❑-❑) and at low ionic strength, before (■-■) and after (▲-▲) shearing, and for "nuclease" chromatin at high ionic strength (Δ-Δ).[11-12-34-35-36] Below: As above, but total scattering intensity.

All the above confirms that *only the long fiber isolated from cold water (with molecular weight larger than 23 Kbp,* Figure 4) *displays a higher order structure at low ionic strenght,* possibly due to a folding back or twisting around each other in a way that was earlier seen by electron microscopy[28-9]

4) *the exact tissue-dependency,* as function of proliferation, differentiation and transformation, of the quaternary and quinternary structure; the latter characterized to a degree previously unmatched by high resolution image analysis of intact nuclei properly stained. Only in recent times other authors became aware of this fact with hydrodinamic studies.

5) *the identification of a structural repeating subunit at higher order, namely the fibrosome*[28] or loop, which can be functionally associated to a gene (Figure 4); the resulting number of about 100.000 genes is compatible with indipendent estimates by classical genetics and recombinant DNA.

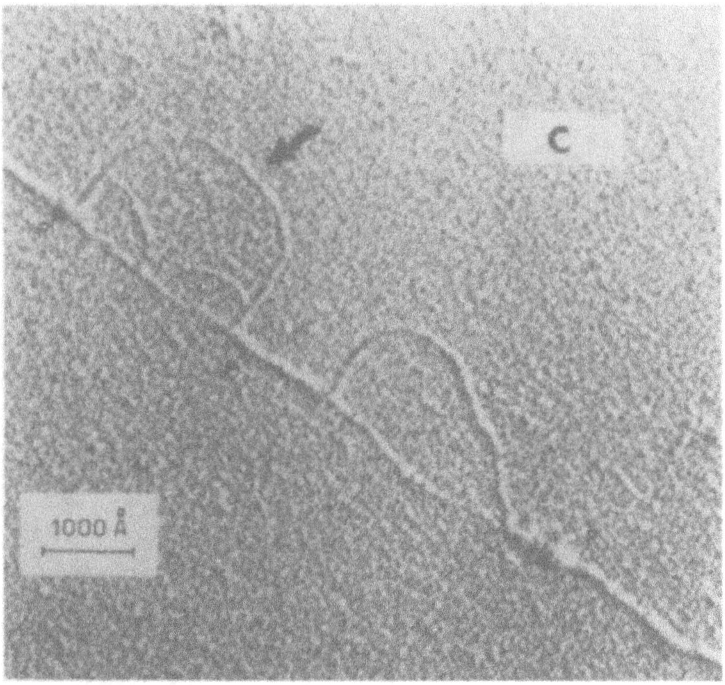

Figure 4. Electron microscopy of native chromatin fiber, displaying highly packed regions of variable lenght alternating with regions of fixed length (called "fibrosomes") unfolding into nucleofilaments[28].

Regardless the exact geometry of the higher order structure, in the chromosomal DNA-protein complexes the DNA is wrapped tightly around a central core of proteins: up to few hundred base pairs (bp) in few superhelical turns. In both nucleosomes and polynucleosomes the diameter of the superhelix is only few times larger than the diameter of the DNA double helix itself. To accomodate curvature of this magnitude all the grooves (both major and minor) on the inside of the curve must narrow considerably, owing to the compression associated with bending, whereas those on the outside of the curve become correspondingly wider. The changes in the base-pair stacking necessary to achieve this will depend on the rotational setting of the DNA relative to the direction of bending.

The formation of nucleosome-like structures is a common feature of processes such as site-specific recombination with the two recombining DNA helices being closey juxtaposed in a precise spatial arrangement. One example of this is the integration of phage l DNA into the bacterial chromosome[48]. This recombination takes place between two attachment sites on the bacterial and phage chromosomes, differing substantially in complexity, one wrapped in a nucleosome-like structure, the intasome, around a protein complex containing the l int protein and the integration host factor. Recombination requires DNA, to enhance the affinity of int protein. However, as it will be shown later, the primary role of supercoiling is to facilitate the formation of a wrapped, functionally active, higher-order structure rather like the nucleosome, the polynucleosome and even its ever higher order superfolding.

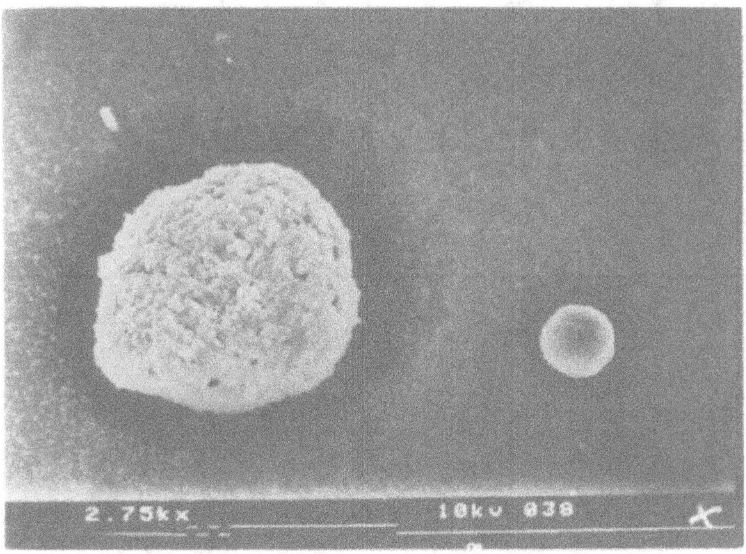

Figure 5. Living mammalian epatocyte (about 14 m diameters) and the corresponding isolated nucleus (about 4 m diameters).

3D INTRANUCLEAR DNA DISTRIBUTION

To prove the in situ 3D DNA distribution living (Fig. 5) and native nuclei (Fig. 6) isolated from animals with collagenase method are typically stained with a fluorescence dye.

Fluorescent nuclear images are then acquired with a 100x 1.3 **NA oil**-immersion objective mounted on a Zeiss Axioplan microscope set-up for **epifluorescence**; successive images are acquired by progressive defocusing, by changing the focal plane in either direction within the object by a small increments of 1.35 mm. The progressive defocusing is performed by a computer driven stepper motor.

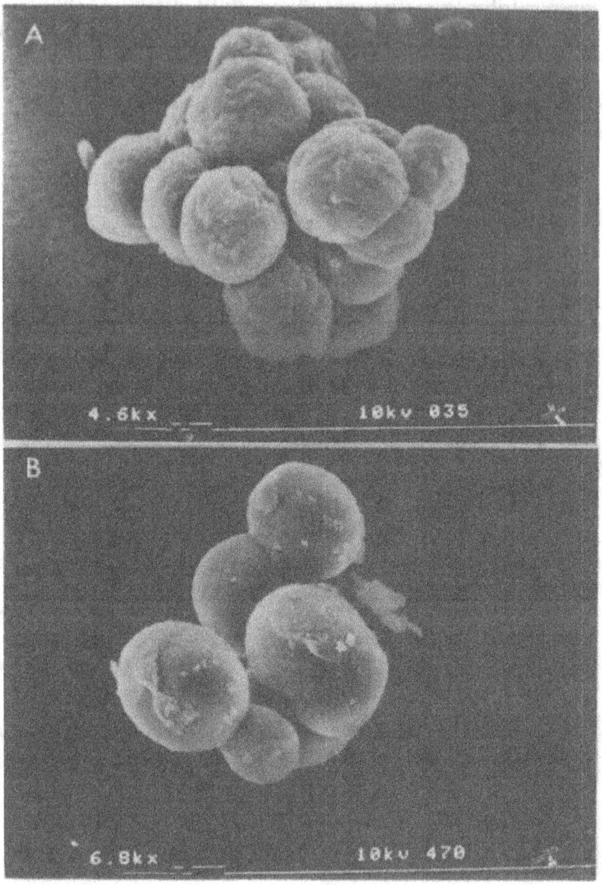

Figure 6. Intact (above) and triton-washed (below) native nuclei (Gavazzo et al., in preparation) isolated from mammalian hepatocytes.

The computer workstation consists of an IBM enhanced AT computer. A DAC/IBM card is used to control the z-axis movements and the shuttering system drivers. To manage the acquisition, process and display operations, an IMAGO software system was used[11-12]. Our goal was to follow dense chromatin zones (paths), with the aid of 3D modelling and cytological mapping, to reveal a number of consistent organizational motifs that characterize their spatial organization typically near the nuclear pariphasy and around the nucleidi (Fig. 7).

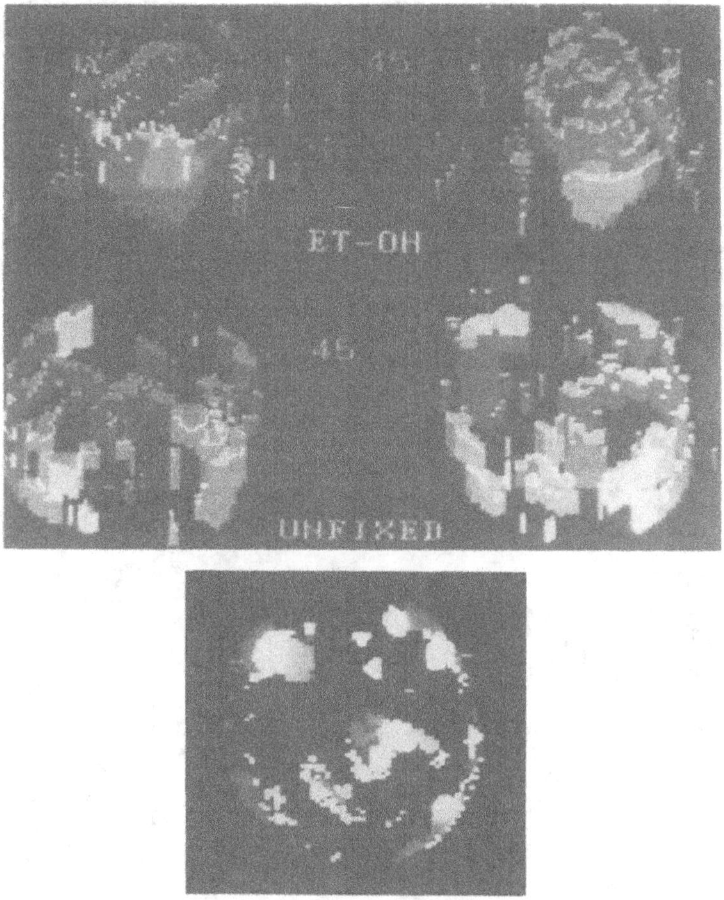

Figure 7. DNA distribution in successive optical sections of native nuclei DAPI-stained.

Functional changes in any mammallian cell are always accompanied (or even preceded) by significant changes in the higher order structure of the cell and of its main constituents, monitored by computer controlled optical sectioning of the object in conjunction with an electromechanic light shuttering system. This sectioning allows us to look closely into the sample without disturbing it. During the last few years there has been a great deal of interest in confocal microscopes, designed specifically to give optical sections. Although confocal microscopes are much better at rejecting information that contaminates section images than is optical sectioning microscopy, confocal microscopes present some disadvantages for our application. The method implemented to remove the out-of-focus information uses a new approximate method, developed in terms of spatial frequency analysis. The whole system has been realized using as main constituents a PC workstation and a modified optical microscope[11-12]. Through this approach higher-order chromatin structural domains approximately 130 nm in width have been recently observed as prominent components of human mitotic chromosomes using buffer conditions which preserve chromosome morphology as determined by light microscope comparison with chromosomes within living cells[3]. Spatially discrete chromatin structural domains of similar size also exist as prominent components within interphase nuclei prepared under equivalent conditions and even in "cold water" native chromatin (see also Fig. 8). Three-dimensional views using stereopairs of chromosomes and interphase nuclei from 0.5 mm thick sections suggest that these 130 nm large-scale chromatin domains consist of 30nm fibers packed by tight folding into larger, linear, fiber-like elements". Reduction in vitro of either polyamine or divalent cation concentrations within two different buffer systems results in a "loss of these large-scale domains, with no higher-order chromatin organization evident above the 20-30 nm fiber". These results suggest that "these large-scale chromatin structural domains might be fundamental elements of chromosome architecture in vivo", namely its *quinternary structure,* which remains even after fixation as long as appropriate cation concentration is present.

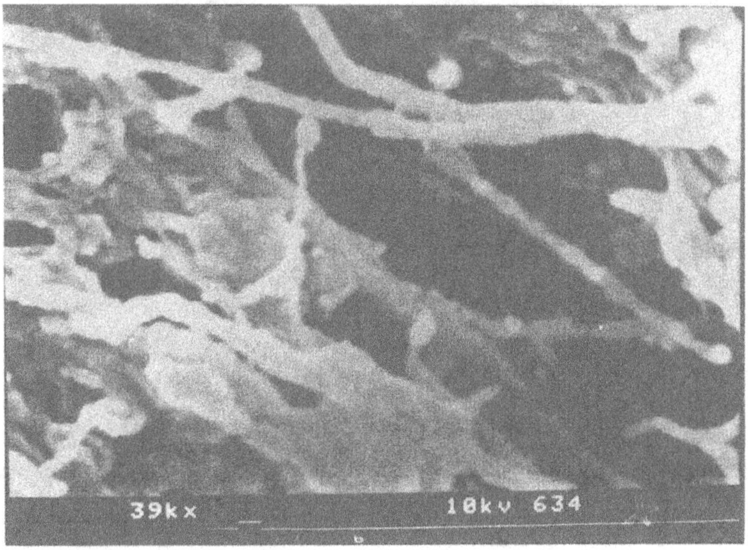

Figure 8. Scanning electron microscopy of interphase chromosome fibres isolated by the "cold water" method (Gavazzo et al., in preparation).

This despite effects of the well known common fixatives on nuclear and cellular size: ethanol induces a dramatic coartation of nucleus only partially affected by acetic acid, while glutaraldheyde gives a network between the amino-groups of proteins. The effects of these fixatives on three-dimensional distribution of chromatin within native nuclei is now under study to prove above results[3].

Recently fluorescence hybridization to interphase nuclei in liquid suspension has shown to allow also quantification of chromosome-specific DNA sequences using flow cytometry and the analysis of the three-dimensional position of these sequences in the nucleus using fluorescence microscopy. The three-dimensional structure of nuclei is substantially intact after fluorescence hybridization in suspension, permitting the study of nuclear organization by optical sectioning. A chromosome-17-specific repetitive probe was used to demonstrate that target sequences as small as one megabase (Mb) can be detected using fluorescence hybridization and flow cytometry[47].

GENE

Recently we have carried out in situ hybridization techniques to monitor oncogene location and selective expression or repression in correlation with 3D intranuclear DNA distribution and local supercoil[49]. The cells (either HeLa or CHO) fixed with ehanol-acetic acid (3:1 V/V). The oncogenes used as probe are labelled by random primed incorporation of digoxigenin-labelled dUTP (deoxyuridine-thriphosphate) for about 12 hours at 37°C. Hybridization at low temperature is accomplished in 25% formamide and 10% dextran sulfate in 2 X SSC containing 7 μg/ml of carrier DNA and 0.4 μg/ml of probe. At last each slide has been mounted with 20 μl of 15 μM Dapi solution in PBS and cells immediately 3D reconstructed by optical microscopy and computer processing.
The hybridization between labeled probe and genomic sequences can be revealed by the presence of dark grains visible by absorbance optical microscopy; the good signal/noise ratio (as shown in figure 2) makes possible to correlate number of grains to the copies of genes inside of the cell genome. Interactive thresholding utilizing look up table facilities has been used in order to obtain a quantitative recognition and a good discrimination of grains. A line plot of a row connected with a grain provides a valuation of its size and gray level. The densitometric profiles so obtained are represented in figure 2.
The in vivo results have been compared with in vitro experiments through hybridisation on nylon filter of DNA extracted from synchronized CHO cells (M, early S-phase) with the same probe used for in situ analysis. Dot blot experiments have been used as semi-quantitative methods for determining the relative levels of oncogene sequences in different samples. Policlonal antibodies for corresponding oncoprotein are used in order to check the expression of the oncogene under sudy directly inside of the cell. The comforting results obtained for mitosis, G1 and early-S phase points to the feasibility and accuracy of in situ simultaneous characterization of single genes structure and of chromatin-DNA structure. Indeed the number of c-myc gene copies being detected appears about twice in 4C cells (mitosis) in comparison with the corresponding 2C cells (synthesis) as shown in figure 9.

HISTONE H1

The lysine-rich chromosomal protein histone H1 has been shown to be critically related to the control of the tertiary (nucleosome) and quarternary (solenoid or rope-like) structure of mammalian chromosomes[28-30-31], and thereby ultimately of gene expression[13].
In solution and under physiological conditions of pH and ionic strength, histone H1 is well known to consists of three structural regions, two random-coil N-terminal and C-terminal regions and a central globular region which is highly structured and conserved. This globular part or "core", which extends from residues 39 (±4) to 116 (±4), with a molecular weight of about 9000 Daltons, has an unknown function, although it has been suggested that it seals the DNA on the nucleosome surface.
Little known at the atomic resolution about the secondary and tertiary structure of the globular domain of H1 histone, and so far no high quality crystals (i.e. diffracting to

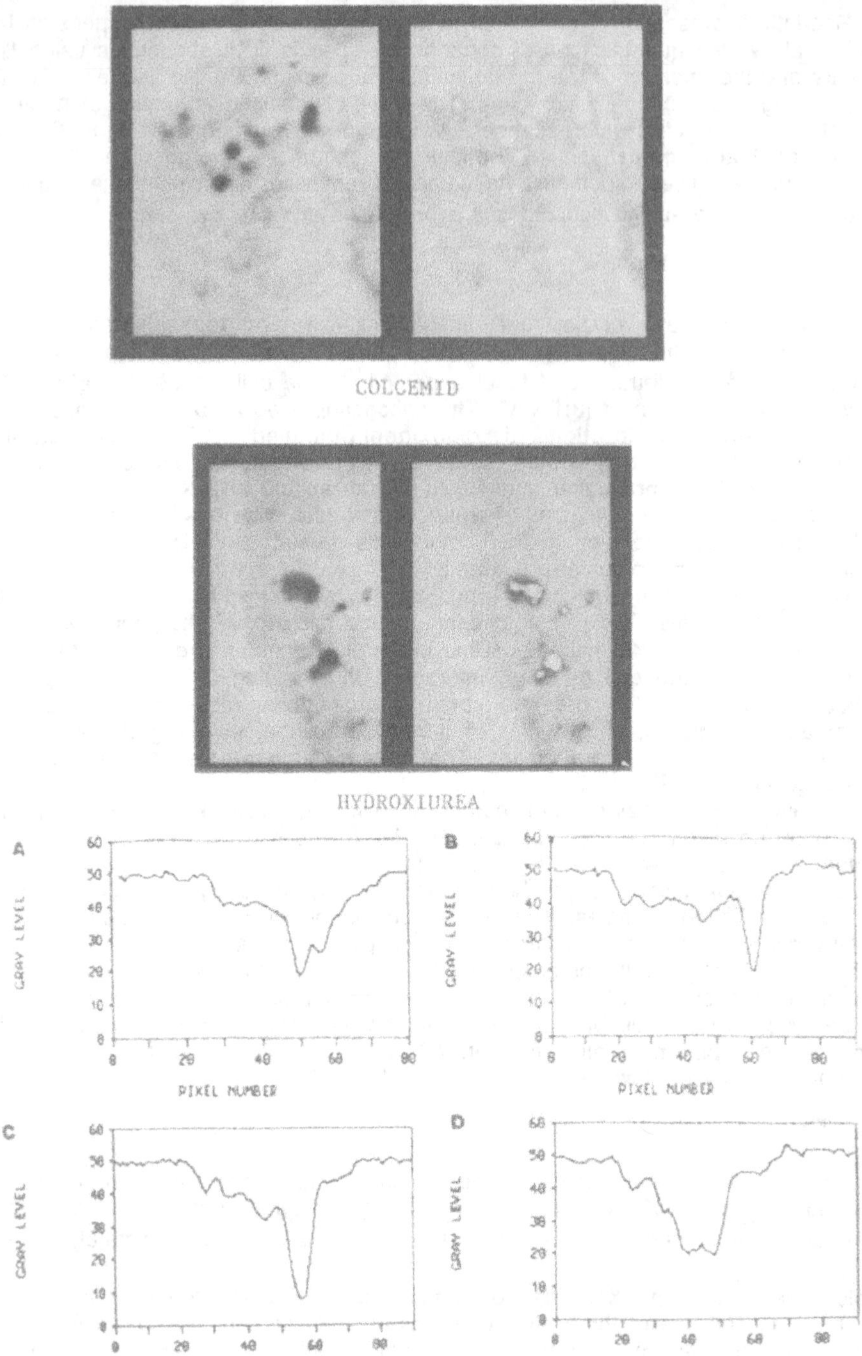

Figure 9. Photo of M (A) versus early S (B) CHO cells after in situ hybridization (A). Grey level as function of pixel number is also shown for one line crossing the grains which are pointed out by image thresholding (C).

better than 5 Å resolution for X-ray crystallographic studies) have been reported. Our present understanding of the structure of the globular part of histone H1 up to the atomic resolution is based on complementary experimental probes[42] and on a combination of statistical methods and molecular modeling. Structural details of the globular domain of H1 histone were obtained by 2DFT protein NMR spectra taken at high resolution[37-38], while its alteration due to changes in PH and temperature were monitored by Circular Dichroism (CD) spectra, differential scanning calorimetry and 1DFT proton NMR spectroscopy[42].
^{31}P and ^{1}H NMR relaxation parameters of nucleosomes isolated from mammalian cells[37-38], as a funcion of the selective progressive removal of the bound chromosomal proteins (HMG,H1, H2A, H2B,H3 and H4), confirm the critical role of H1 histone core in the control of DNA internal motion within nucleosomes.
From the experimental CD, NMR, calorimetric and theorical studies, the histone H1 core appears to be consistently structured. This despite the fact that in the 2DNMR experiments the need to operate at relatively low pH (to reduce proton exchange in water) has probably caused the loss of some structure. The consequent overlap of many peaks also hinders the complete assignments of the resonances. However, observation of the short and medium range NOEs allows us to say that a significant part of the protein is still structured and to obtain a possible secondary structure in agreement with the CD and statistical results (Table 1).

The amount of a-helical secondary structure estimated from NMR data is identical to the earlier CD estimate (30%) of the same H1 core, but slightly less than generated by the CD measurements reported here which have been corrected for inactive sample artefacts. The assignments and NOE data are poorer in the second half of the primary sequence, and the last two of the four helices proposed could be bigger. On the basis of long range NOE data we can at present propose the 3D folding shown in figure 10 for the H1 histone globular region.

Earlier nuclear magnetic resonance studies have shown that internal structure of B-DNA experiences large fluctations in nucleotide conformation which are strongly affected by the binding of ethidium bromide. Similarly, ^{31}P (Table 2) and ^{1}H NMR spectra of trinucleosomes suggest that as a result of chromosomal histone proteins binding nucleosomal DNA internal motions are greatly hindered. Indeed, we show that the selective removal of chromosomal proteins by progressively increasing the concentration up to 2 M NaCl, progressively increases the internal motions up to that of free DNA[39-40].

The H1 histone extraction at 0.6 M NaCl, as compared to the selective extration of HMG non histone (0.35 M NaCl) and H2A-H2B histone (1.2 M NaCl), displays the largest variation of ^{31}P line width (Table 2).

This is compatible with earlier suggestions on its role in the sealing of nucleosomes.
One of the features of the interaction between histone H1 and naked DNA which has attracted considerable attention is the preference of H1 for supercoiled DNA forms. Recently authors investigate the binding of H1 to superhelical and linear DNA forms by the use of direct competition experiments. In addition, they compared the behavior of H1°, the differentiation-specific subfraction of H1, with that of H1. The idea behind this comparison was to see whether the different mechanism of action of the two histone H1 subtypes might be somehow connected to the way they interact with differently supercoiled DNA The answer was negative[55].

Table 1. Secondary structure of the histone H1 core from 2DFT-NMR

```
 35                  40                  45                  50                  55
--Lys-Ala-Ser-Gly-Pro-Pro-Val-Ser-Glu-Leu-Ile-Thr-Lys-Ala-Val-Ala-Ala-Ser-Lys-
Glu-Arg-
  C   C   C   C   C   T   T   T   T   B   B   B   B   B   B   A   A   A   A   A   A
                 60                  65                  70                  75
--Ser-Gly-Val-Ser-Leu-Ala-Ala-Leu-Lys-Lys-Ala-Leu-Ala-Ala-Ala-Gly-Tyr-Asp-Val-
Glu-Lys-
  A   A   C   C   C   C   C   C   C   C   C   C   C   C   C   C   C   C   C   C   A
             80                  85                  90                  95
--Asn-Asn-Ser-Arg-Ile-Lys-Leu-Gly-Leu-Lys-Ser-Leu-Val-Ser-Lys-Gly-Thr-Leu-
Val-Glu-Thr-
  A   A   A   A   A   A   C   C   C   C   A   A   A   A   A   C   C   C   C
         100                 105                 110                 115
--Lys-Gly-Thr-Gly-Ala-Ser-Gly-Ser-Phe-Lys-Leu-Asn-Lys-Lys-Ala-Ala-Ser-Gly-
Glu-Ala-Lys-
  C   T   T   C   C   C   C   A   A   A   A   A   C   C   C   C   C   C   T   T   T
     120
--Pro-Lys.
  T   C
```

Comparison between the different amounts of secondary structure calculated from NMR and CD approaches. The results shown here are reported from the 4 helices model at pH 4.4 in Table 5.

	NMR	CD
Åa-helix:	26 res (30%)	35 res (41%)
b-sheet:	6 res (7%)	- - -
turn:	10 res (12%)	2 res (3%)
random coil:	44 res (51%)	39 res (56%)

A = a-helix; B = b-strand; T = turn; C = random coil.

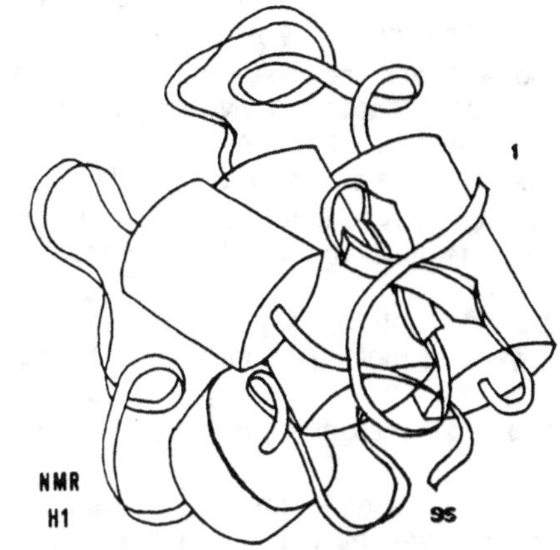

Figure 10. Preliminary model of H1 histone care tertiary structure at atomic resolution, from 2DFT NMR, CD and DSC date[37-38-42].

Table 2. Trinucleosome ^{31}P. Spectra half width - half height.

10 nM NaCl	525 Hz
0.35 M NaCl	465 Hz
0.6 M NaCl	404 Hz
1.2 M NaCl	404 Hz
2.0 M NaCl	323 Hz

CHROMATIN STRUCTURE AND REDUCED ACCESSIBILY TO DNA OF SEQUENCE-SPECIFIC PROTEINS

One of the key role of higher order chromatin structure and super folding appears to be the repression of the initiation of gene transcription, by several mechanism:
1) the alteration of base sequence-dependent DNA secondary structure by nucleosome formation.
2) the H1-nucleosome association which stabilizes the high order 30 nm chromatin fiber[13-28-6] making thereby "invisible" the DNA to sequence specific proteins;
3) the active repression of certain genes by histones (namely H1 and H4) through the displacement of transcriptional factor, as shown in Xenopus class III[54];
This repression in turn could be bypassed in vivo respectively by:
1) rearranging a relevant surface of DNA helix for exposure to transcription factors, such as TF III A, capable to recognize bent non form-B DNA.
2) having a reduced amount of DNA to search of its binding sites[22] to H1, which bind across the dyed axis of nucleosomes and stabilize peripheral 20 bp DNA entering and exiting around the octamer core;
3) making, through fiber disruption, highly visible to RNA polymerase each preinitiation complex; in general molecular mechanisms unravelling the 30 mm fiber would enhance the transcription;
4) enzymatic modification of histones, namely H1 phophorilation[13] and H4 acetylation[1];
5) nucleosome displacement, as consequence of transcriptional gene activation, by associating transcription factor over the PH05 promoter region nuclease-accessible.
The exact molecular description and cause-effect relationship are still lacking howerver. One issue still unsolved is the existence or absence of a precise positioning of nucleosomes relative to a DNA sequence being warranted by a specific feature of DNA secondary structure or by certain "phasing" proteins.
Nucleoplasmin, or other nucleosome assembly proteins, may mediate nucleosome dissociation during activation; another "window" could be the exact time during the cell cycle at which all DNA is likely to be accessible to transcriptional factor during early S; chromatin takes indeed several minutes to "mature" following replication fork passage.

INTRANUCLEAR DNA DISTRIBUTION AND DNA REPLICATION-TRANSCRIPTION

Using HeLa S-3 cells synchronized by selective detachment, we[30-31] have previously reported a parallel study of nuclear morphology and autoradiography grain patterns between middle GI and middle S phases. Our results show two distinct [H]-thymidine labelling patterns. The first "peripheral" labelling pattern has a characteristic nuclear size distribution, in contrast to the heterogeneous and varying size distributions of Feulgen-stained nuclei, and apparently is characteristic of very early S phase. The size of the second labelling pattern - homogeneous or inhomogeneous grain distribution throughout the nucleus - are equal or larger extreme of nuclear areas, and the labelling index closely parallels the fraction of nuclei with areas larger than the minimum size of the labelled nuclei. These results suggest a characteristic nuclear size (reflecting unique intranuclear DNA distribution) as a necessary, if not sufficient, requirement fo S phase initiation. Parallel experimentation with rat liver cells - synchronized in vivo by partial hepatectomy and analyzed by thin section autoradiography - confirm the existence of a peripheral labelling pattern in both the very late and very early part of S phase, which reconciles our data with previous results and points to the fact that both initiation and determination sites for DNA replication are near the nuclear periphery.

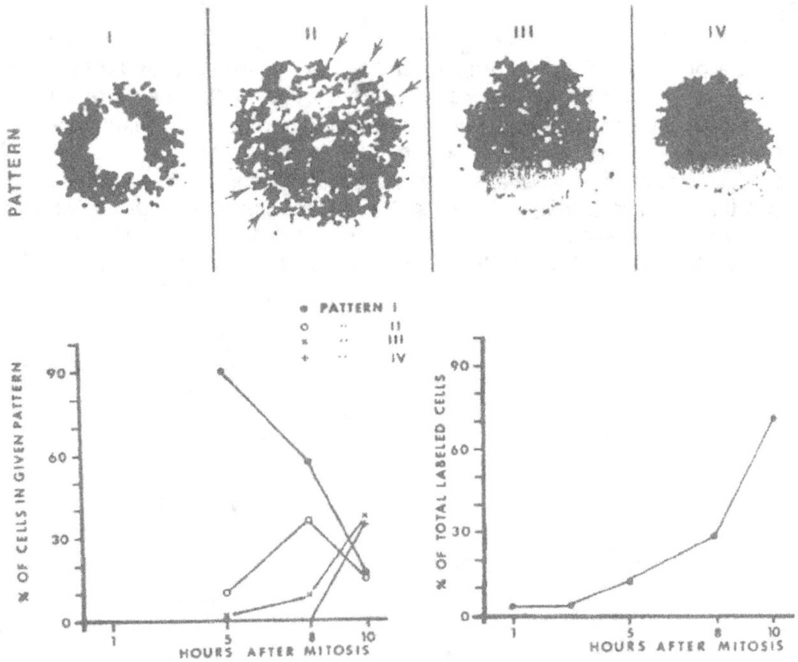

Figure 11. Peripheral labeling pattern, typical of very early and very late S phase. Its correlation with unique nuclear size and DNA synthesis initiation[30-31].

More recently[33-34-35-36] mammalian cells and nuclei were shortly digested with either micrococcal nuclease or DNAse I, both before and after mild fixation either before (G0) or after (G1) partial hepatectomy. Cells were Feulgen stained and examined by high resolution light microscopy. In metabolically active G1 nuclei, intranuclear DNA appears organize at least in two distinct domains, whereby the highly dispersed one is large enough to be detected at the resolution of the light microscope and appears preferentially attacked by limited DNAse I digestion. The action of the enzyme is readily apparent only in the nuclei that are first digested and then fixed. Spectroscopic characterization of the same nuclei reveals that the fixation causes a sizeable removal of proteins, mostly in the soluble chromatin subfraction. These results have been explained in terms of two distinct control levels for gene expression and for higher order DNA structure, with the form Z being associated to loosely bound HMG and H1 histones and the latter being associated to the class of strongly bound non-histone (laminar) proteins forming a network apparent in metaphase chromosomes and interphase nuclei after exposure to DNA-se, RNA-se, detergent and 2M NaCl[33-34-35-36].

The state of supercoiling of intracellular DNA in eukaryotic cells has been subject to controversy, and contradictory views are often found in the literature. However, it has been conclusively shown[16] that DNA supercoiling can occur in eukaryotes. Furthermore several lines of evidence point to supercoiled DNA as more highly expressed than linear DNA, whenever introduced into living cells[52]; quite compatibly with the fact that *in vitro transcription of eukaryotic genes is strongly affected by the degree of DNA supercoiling.* Several genes are optimally transcribed under various extents of supercoiling, suggesting that DNA topology plays a role in the regulation of gene expression[18].

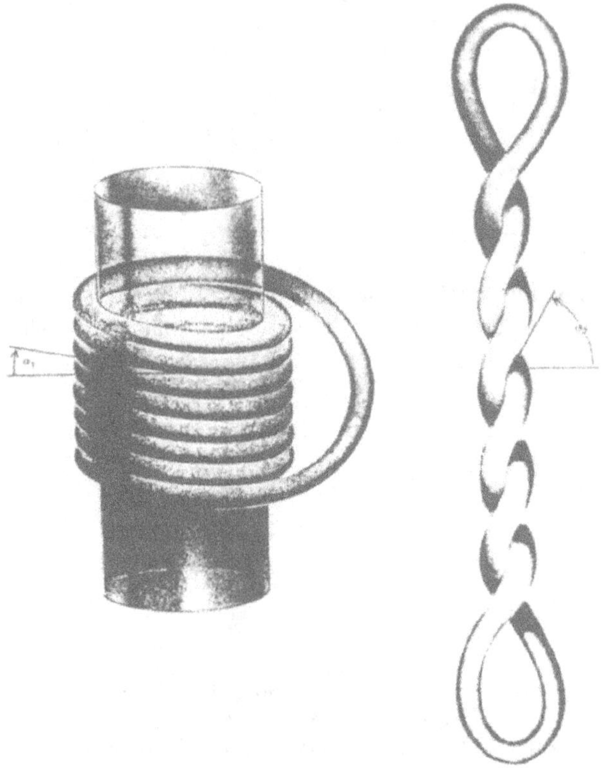

Figure 12. Rubber like induction of negative superhelical density in "closed" DNA like the one present in polynucleosomed fiber mantained at fixed sites in native nuclei.

The supercoil-induced B to Z transition in recombinant plasmids has been the subject of several investigations in the recent past. The structural inversion of a left-handed helix form was observed in segments of alternating dC-dG. These structural and energetic alterations demonstrate that methylation of cytosine in the 5-position may be an important switch mechanism for influencing the B-Z equilibrium and DNA topology in general, thus potentially affecting DNA-protein interaction and gene regulation at physiological levels of DNA supercoiling (Figure. 12).

Interestingly, transcription of a DNA molecule inside a bacterium *is accompanied by* local and temporal *supercoiling* of the DNA template: as transcription proceeds, DNA in front of the transcription ensemble becomes positively supercoiled, and DNA behind the ensemble becomes negatively supercoiled. Because bacterial gyrase and topoisomerase I act differently on positively and negatively supercoiled DNA, the formation of twin supercoiled domains during transcription is manifested by a large increase or decrease in the linking number of a intracellular plasmid when bacterial DNA gyrase or topoisomerase I, respectively, is inhibited. Such changes in linking number and in the state of DNA supercoiling appear strongly modulated by transcription[18].

The DNA supercoil has been gaining over the last several years much attention as the key structural feature most important in the control of gene expression. Its formation and dynamics appear modulated in solution as DNA interacts with chromosomal proteins or other cellular component. The feasibility to monitor changes in the degree of supercoil is however hampered by the invasiveness and limitations of present technologies, either in situ (electron microscopy) or in vitro (sedimentation). We have then explored with success the utilization of a recently implement technique based on the scattering of circularly polarized light in the characterization of circularized λ-DNA in presence of increasing ethidium bromide concentration.

Typically DNA topoisomerase mutants are used to study the topological state of intracellular DNA, namely switching off the gene topA encoding DNA topoisomerase I leads at the same time to an increase in the degree of negative supercoiling of intracellular DNA, to inhibition of the growth of the cells, and to change secondary structure from a right-handed B-helical structure to a left-handed Z_helical[16].

If the linking number a of a DNA differs from that of the same DNA in its relaxed state, the molecole is under strain. This often results in the coiling of the helical molecule in space, and the DNA is said to be supercoiled. In a supercoiled DNA, there are both torsional and flexural strains. Supercoiling of a DNA therefore has strong effects both on structural changes in a DNA and on interactions between DNA and other molecules like histone H1 (see earlier).

The degree of supercoiling of intracellular DNA in various organisms has been reviewed recently. The values of superhelical density σ per 10 bp for DNA rings isolated from eukaryotes are typically around -0.06. Because of interactions between intracellular DNA and components in the cellular milieu, however, there is some uncertainty in defining the supercoiled state of intracellular DNA. For DNAs in eukaryotic cells, the negative values of σ can be attributed almost entirely to nucleosome formation which lowers the linking number of a relaxed DNA relative to that of the same DNA when relaxed in its pure form. The binding of cellular components appears to reduce the linking number of the relaxed DNA and the effective degrees of supercoiling of intracellular DNA are probably a factor of 2-3 lower than the specific linking differences with respect to pure DNA in its relaxed form as the reference state.

Recent analysis has pointed to the importance of localized supercoiling by proteins translocating along a DNA, and in particular by the transcription process. Because of the high degree of supercoiling that can be achieved by such processes, the range in s that can drive DNA structural transitions in vivo may be much larger than previously believed. A corollary of localized supercoiling is that the accessibility of regions of DNA to DNA topoisomerases may have profound effects on the supercoiling of these regions. Thus, DNA structural changes, DNA supercoiling, DNA topoisomerases, processes involving proteins moving along DNA, and DNA transcription are all intricately connected inside a cell.

DECOUPLING OF NUCLEAR TO CELLULAR GEOMETRY

It has been shown that a higher-order chromatin superpacking and a reduced template activity are associated with the expression of both the transformed phenotype[19-26-27] and cellular senescence[26-27]. This apparent paradox is resolved by measuring the degree of coupling between changes in nuclear and cellular morphometry. As cells (Figure 5) pass through their cycles, the correlation between the nuclear and the cellular form factors is high for young fybroblastlike cells (Figure 13), and very low for transformed cells. This correlation is increased but still poor in tumor cells with lower metastatic potential[21], but becomes nearly perfect in old fibroblasts[26-27]. The mechanisms by which cell geometry and growth are coupled in normal cells and uncoupled in transformed cells[15] could indeed be the physically (microtubules-microfilaments) or chemically induced coupling between cellular and nuclear morphometry[26].

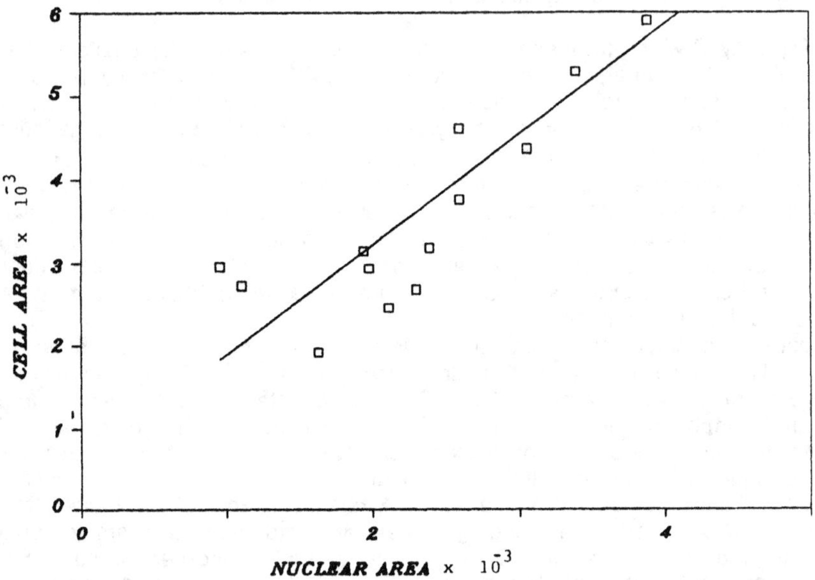

Figure 13. Nuclear versus cellular area in normal fibroblast like cells, monitored by high resolution image analysis following double-stain with DAPI and fluorescience-labeled anti-tubulin monoclonal antibodies[49-50];

MICROTUBULES DISORGANIZATION

When cells are stained with antitubulin antibody and examined by indirect immunofluorescence, they display an elaborate coplex of fine fluorescent filaments throughout the cytoplasm, radiating from two or more specific foci near the nucleus out toward the cell periphery, and terminating at the plasma membrane (Figure 14). Recent attention has been focused on the cytoplasmic microtubule complex (CMTC), and on the correlated intermediate filaments and microfilaments, because of their disappareance or disorganization in transformed cells. The CMTC is present in normal cells in all phases of the cell cycle, except at the beginning of mitosis, where it is disassembled and replaced by a highly fluorescent mitotic apparatus that functions to segregate the chromosomes during mitosis.

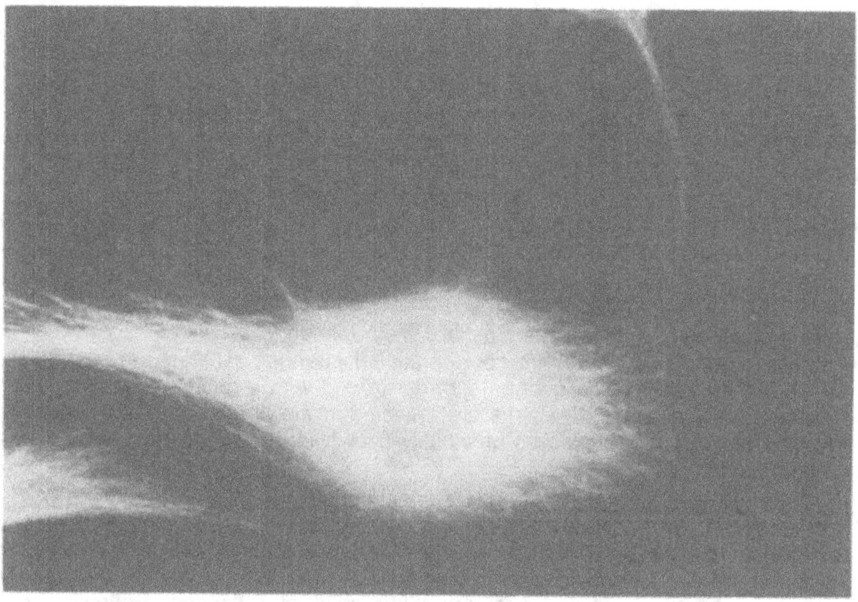

Figure 14. Microtubules network of mammalian cells stained with fluorescin-labeled anti-tubulin monoclonal antibodies.

Opposing evidence and frequent objections[41] have been raised over the claim that cells that are transformed either by viruses, chemicals, or spontaneously contain fewer microtubules, and that the microtubules are more randomly oriented and shorter in lenght than their normal counterparts[43].

What counts is the geometric location of the microtubules, not their total amount. The failure of microtubules to reach the cell membrane could by itself explain the observed lack of coupling between cell geometry and chromatin structure and function in transformed cells. In normal cells, however, the presence and orientation of microtubules can constitute a further independent constraint to cell cycle progression, which induce changes in chromatin structure and function through mere changes in cell geometry, as is done with poly-HEMA[3-15], where the metabolic activity of fibroblast may be decreased down to a halt, even when all serum growth factors and chemical constituents remain present and unchanged. Conversely, changes at the chromatin and nuclear level may propagate to the same mechanism through the cell, causing changes in membrane permeability and ion transport, which in turn may affect chromatin and normal cell cycle progression. Thus, the preservation of the microtubule's complex integrity and attachment is an initial step for normal cell expression and control of human cancer.

WATER AND IONS

The complex dielectric constant of small quantities of liver nuclei in various functional states was measured in the frequency range of 50-2,000 MHz using an Automatic Network Analyzer. From these measurements, through an electric model of macromolecules in solution, several quantities such as ion content, bound water, and free water have been estimated. Unique changes in the physical state of intranuclear water and ions were then apparent in the resting liver nuclei immediately following induced cell proliferation, as compared to nuclei either from early carcinogen-altered hepatocytes or from late selected carcinogen-initiated hepatocytes. Possible implications of these findings become apparent in terms of the molecular events controlling chemically induced neoplastic transformation[32].

DNA strand breaks and chromatin conformational changes (15) are among the early alterations induced in liver cells when a rat is treated with a chemical carcinogen. It is then not surprise that such alterations are reflected in the amount and physical state of bound water in the intact nuclei, as previously shown for serum proteins in normal and pathological conditions.

More recently[33-34-35-36] proton-NMR spectroscopy of liver nuclei of 1, 3, 6, 12 and 24 month old rats reveals a significant increase in the T1 relaxation time with increasing age, compatible with the observed increase in chromatin-DNA condensation and in the fraction of free for a constant amount of total water. Little or no increase in T1 takes instead place in the corresponding cells and tissues. Following partial hepatectomy, both the cell properties - in terms of either tritiated thymidine or methachromatic dye uptake - and the exponential decay of total magnetization - at every level - appear bimodal. The splitting into two T1 subpopulations (G0 and G1) becomes less pronounced with age, in agreement with the age-dependent decrease in the fraction of cells responding to the proliferative stimulus. It appears worthy of notice the T1 value of G1 nuclei which is significantly lower than the corresponding G0 nuclei for a constant total water content. These data are compared with the T1 values obtained for preneoplastic and neoplastic populations chemically induced in the same rat liver model, displaying high correlation with other biophysical parameters and some clinical significance[33-34-35-36].

ONCOGENES AND THEIR UNBALANCED EXPRESSION

Abnormal expression of gene or of several proto-oncogenes may cause the beginning of human cancer.

Among the nuclear proteins coded by proto-oncogenes (c-onc) are c-Myc, c-Foc, and c-Jun. These proteins have the potential to regulate other genes by interacting directly or indirectly with DNA, thereby influencing gene transcription. The c-fos gene encodes phosphoprotein with a half-life of approximately 20 min. The c-myc gene encodes a nuclear phosphoprotein of 58-66kD and similar to the c-Fos and c-Jun proteins it has a short half-life. The c-fos, c-jun, and c-myc genes are activated when quiescient cells are stimulated to proliferate by the addition of growth factors.

As previously shown[30] and earlier discussed, the end result of all induced alterations, regadless of the molecular mechanism, appears to be a local change at the level of single fibrosome unit, namely on its number of negative superhelical turns per unit DNA. It is likely through this mechanism of induced high twisting of the DNA molecule, that the cell develops a type of energy storage, which is then transferred during normal and abnormal transcription. Indeed, cell posses enzymes, such as topoisomerase or gyrase which decrease or increase the linking number of each fibrosome, depending on the state of chromatin-DNA structure prior to its interaction with the enzyme.
Upon certain conditions, as those created by the overall increase in chromatin condensation occurring during cancer-induction, followed by a strand nicking and closing with topoisomerase, super-superhelical twists are created in individual fibrosome yielding a stored energy above the threshold level, which could then trigger the transcription of the corresponding DNA sequences also through an enhanced binding of RNA polymerases. At the same time, similar condictions, depending on a pre-existing relaxed chromatin structure -i.e., for active genes- at the level of a single fibrosome, could induce a repression of the corresponding genes (Figure 15).

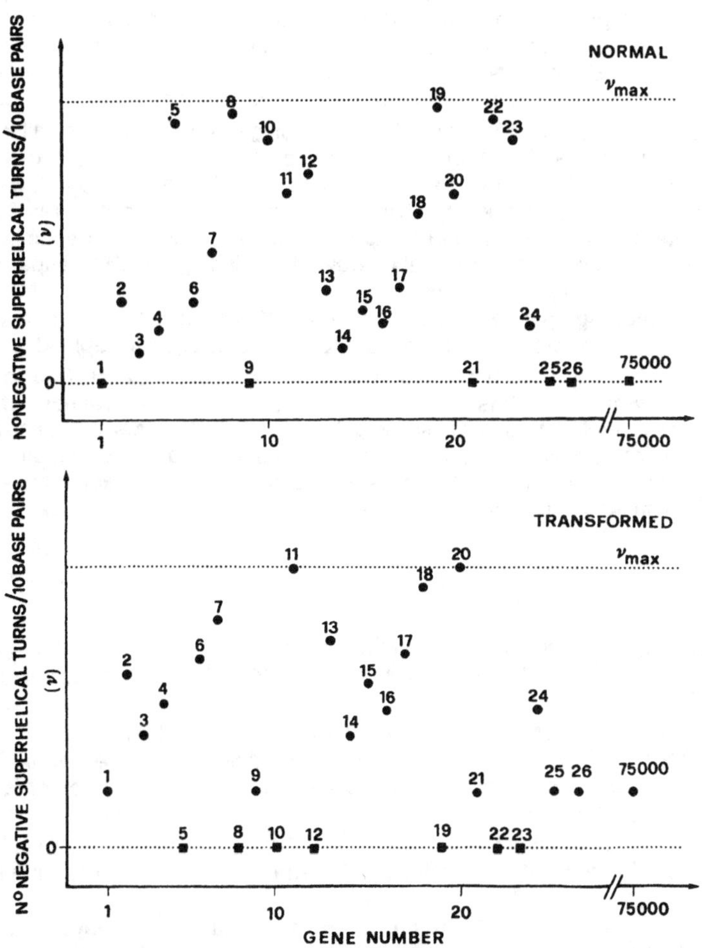

Figure 15. Model for unbalanced gene expression following neoplastic transformation. Number of superhelices per 10 bp for normal (above) versus transformed (below) G1 mammalian cell.

All of the above reconciles the global increase in chromatin condensation to the differential expression and/or repression of the various single genes monitored by genetic engineering in cells which undergo neoplastic transformation. This differential gene expression, in association with alterations at the level of global chromatin-DNA structure and of the cytoskeleton, may play the key role among all molecular events in the control of chemically-induced cell transformation.

Worthy of notice is also the complex, three-dimensional motion of chromatin within the interphase nucleus, with nucleoli and DAPI-stained, fluorescent chromatin domains describing curvilinear trajectories extending throughout the nucleus. Based upon the evidence that the rate of this motion changes with metabolic demands, it may be postulated that also Nuclear Rotation (NR) plays a role in gene expression, by transposing chromatin domains to be transcribed to specific nuclear compartments.

CONCLUSIONS

Among the numerous mechanisms involved in the control of gene expression, the higher order chromatin and gene structure has been shown to be the most critical. Several experiments appear to confirm the existence of a cause-effect relationship between abrupt alteration in chromatin structure and changes in cell function, whereby a high degree of chromatin-DNA supercoil and microtubules disorganization have been proposed as the two main level of control of human cancer. The negative superhelicity of chromatin-DNA appears maintained only in native long chromatine fibers at the level of the single structural units, called fibrosomes or loops, which exist in situ, and can be readily associated to the single functional units called genes[28]. Several competing molecular events, all occurring at the level of single cell during cell proliferation and transformation, have been associated over the last few years with alterations of the number in the negative superhelices and in gene expression; only the in situ characterization now in progress appears however capable to establish a cause-effect relationship among any of the above events including the DNA supercoil and the molecular initiation of human cancer.

It is finally comforting to see that the paradoxical repression of human genome as prerequesite for the expression of the transformed phenotype is upheld by numerous evidences, including recent extensive experimentation in NIH 3T3 cell line[45]. Neoplastic transformation has indeed shown to occur in a fraction of an entire population that is undergoing a physiological adaptation to moderate constraints on its growth; namely late passage cells with more condensed repressed chromatin were about 100 times more sensitive than early-passage cells to transformation (namely massive foci formation) in low concentration of serum.

REFERENCES

1. Alfrey, V., Faulkner, R., and Mirsky, A., 1964, Acetylation and methylation of histones and their possible role in RNA synthesis regulation, Proc. Natl. Acad. Sci.USA. 51, 786.
2. Athey, B., Smith, M., Rankert, D., Williams, S. and Langmore, J., 1990, The Diameters of frozen - hydrated chromatin fibers increases with DNA linker length: evidence in support of variable diameter models for chromatin, J. Cell Biology, 795:806, vol. III.
3. Belmont, A., Bramfeld, M., Sedat, J. and Agard, A., 1989, Laser-scal chromatin structural domain within mitotic and interphase chromosomes in vivo and in vitro, Chromosoma, 98, 129:43.
4. Bordas, J., Perez-Grau, L., Koch, M. H. J., Vega, M. C., Nave, C., 1986, The superstructure of chromatin and its condensation mechanism. I. Synchrotron radiation X-ray scattering results, Europ Biophys J. 13: 157-173.

5. Bordas, J., Perez-Grau, L., Koch, M. H. J., Vega, M. C., Nave, C., 1986, The superstructure of chromatin and its condensation mechanism. II. Theoretical analysis of the X-ray scattering patterns and model calculations, Europ Biophys J. 13: 175-185.
6. Bradbury, E.M., and Nicolini, C., (Eds.), 1986, "NMR in the life sciences", Plenum Press, New York.
7. Brahms, J., Dargouge, S., Brahms, Y. and Wagner, V., 1985, Activation and inhibition of transcription by supercoiling, J. Mol. Biol., 181 (4), 455:466.
8. Brinkley, B.R., 1982, The cytoskeleton: an intermediate in the expression of the transformed phenotype in malignant cells in "Chemical Carcinogenesis" (Nicolini Ed.), Plenum Press, New York.
9. Cavazza, B., Trefiletti, V., Pioli, F., Ricci, E., and Patrone, E., 1983, Higher order structure of chromatin from resting cells. Electron michroscopy of chromatin from calf thymus, J. Cell. Sci., 62, 81.
10. De Boni, U., 1988, Chromatin Motion in interphase nuclei and its potential role in gene expression, Anticancer Research, 8, 885:898.
11. Diaspro, A., Sartore, M., Adami, M. and Nicolini, C., 1990, 3D representation of biostructures Part A and B , Image and Vision Computing, 8, 130:141.
12. Diaspro, A., Bertolotto, M., Vergani, L., and Nicolini, C., Polarized Light Scattering of nucleosomes and polynucleosomes. In situ and in vitro studies,.1990, IEEE Transactions in Biomedical Engineering .
13. Dolby, T.W., Belmont, A., Borun, T.W., and Nicolini, C., 1981, DNA replication chromatin structure and histone phosphorylation and stage specific nonhistone chromosomal proteins, Biochemistri, 18, 1333.
14. Finch, J.T. and Klug, A., 1976, Solenoidal model for superstructure in chromatin, Proc. Natl. Acad. Sci. USA, 73, 1897.
15. Folkman, J., and Moscona, A., 1978, Role of cell shape in growth control, Nature, 273, 345.
16. Giaever, G. and J. Wang, 1988, Supercoiling of intracellular DNA can occur in eukaryotic cells Cell, 55, 849:856.
17. Gurri V. Giaever, Carry, Snyder, L. and James Wang,, 1988, DNA Supercoiling in vivo", Biophysical Chemistry, 29, 7:15.
18. Hirose, S. and Suzuki, Y., 1988, In vitro transcription of eukaryatic genes is affected differently by the degree of DNA supercoiling, P.N.A.S. - USA 85, 718:722.
19. Kendall, F., Beltrame, F., and Nicolini, C., 1979, Chromatin study in situ. Image analysis in "Chromatin structure and function" (Nicolini, C. Ed.), Plenum Press, New York.
20. Koch, M., Sayers, Z., Verge, M. and Michon, A., 1987, "The superstructure of chromatin and its condensation mechanicm", Europ. Biophy J., 15, 133:140.
21. Lessin, S.R., Abraham, S.R., and Nicolini, C., 1982, Byophisical identification and sorting of high metastatic variants from B16 melanoma tumor, Cytometry, 2, 407.
22. Lin, S., and Riggs, A.D., 1976, "The binding of lac repressor and the catabolite gene activator protein to halogen-substitute analogues of poly (d (A-T)), Biochem. Biophys Acta. 432, 185.
23. Nicolini, C., Baserga, R., and Kendall, F., 1976, DNA structure in a sheared and unsheared chromatin, Science, 192, 796.
24. Nicolini, C., and Kendall, F,.1977, Differential Light Scattering in CD spectra of chromatin: alternative corrections and inferences, Phisiol. Chem. Phis., 9, 265.13.
25. Nicolini, C., (Ed.) 1979, Chromatine structure and function, Plenum Press, New York.
26. Nicolini, C., and Beltrame, F., 1982, Coupling of chromatin structure to cell geometry during the cell cycle: transformed versus reverse-transformed CHO, Cell. Biol. Int. Rep., 6, 63.

27. Nicolini, C., Grattarola, M., Beltrame, F., and Kendall, F., 1982, Coupling of nuclear morphometry to celle geometry: its role in the controll of normal and abnormal cell growth in "Cell Growth", (Nicolini, C., Ed.), Plenum Press, New York.
28. Nicolini, C., 1983, Chromatin Structure: from nuclei to genes, Anticancer Research, 3, 63:86.
29. Nicolini, C., 1984, Nuclease structure and hgher order gene structure, Toxicologic Path, 12, 149:154.
30. Nicolini, C., 1986, "Biophysics and Cancer", Plenum Pubbl. Co., New York, pp. 1:463. 211:218 (1990).
31. Nicolini, C., Belmont, A. and Martelli, A., 1986, Critical nuclear DNA size, Cell Biophysics, 8, 103:117.
32. Nicolini, C., Carlo, P. and Ridella, S., 1987, The physical state of intranuclear water and ions , Toxicologic Pathology, 15, 184:9.
33. Nicolini, C., 1988, Nuclear architecture and three-dimensional organization of chromatin: their role in the control of cell fuction in "Heterochromatin", ed R.S. Verna, Cambridge University Press, Cambridge, 228:249.
34. Nicolini, C., Diaspro, A., Vergani, L. and Bertolotto, M. , 1988, Nuclear architecture, DNA distribution nucleosome and digestion, Cell Biophysics, 13, 1:14.
35. Nicolini, C., Diaspro, A., Vergani, L. and Cittadini, G., 1988, In situ thermodynamic characterization of chromatin and of other macromolecules during cell cucle, Int. J. Biol. Macromol.,10, 137:144.
36. Nicolini, C., Vergani, L., Diaspro, A., 1988, Native chromatin and DNA demage induced by nuclease, Bioph. Bioch. Res. Com., 155, 1396:1403.
37. Nicolini, C., Catasti, P., Nizzari, M., and Carrara, E., 1989, "Histone H1 solution structure and the sealing of mammalian nuclei" in Protein Structure et Engineering eds. Nicolini C. and Jardetzky O., Plenum Publ. Co., 220:241.
38. Nicolini, C., Martelli, A, Zaccheo, D., Robbiano, R., 1989, Aging, Differentation and proliferation: a proton NMR study, Magnetic Resonance in Medicine and Biology, 253:70.
39. Nicolini, C., Diaspro, A., Facci, P., Bertolotto, M. and Vergani, L., Changes in DNA superhelical density monitored by CIDS, Eur. Biophys. J., in press.
40. Nicolini, C., Diaspro, A., Facci, P., Bertolotto, M. and Vergani, L., Changes in DNA superhelical density monitored by CIDS, Eur. Biophys. J., in press.
41. Osborn, M., and Weber, K., 1977, The display of microtubules in transformed cells, Cell, 12, 561.
42. Pepe, M., Catasti, P., Rauch, J., Nizzari, M., Nicolini, C., 1990, The histone H1 globular region. A supersecondary structure from spectroscopic and statistical studies, 1990 Biochem Biophys Acta, 1041, 14-21.
43. Puck, T.T., Waldren, C.A., and Hsie, A.W., 1972, Membrane dynamics in the action of dibutyryll cAMP testosterone on mammalian cells, Proc. Natl. Acad Sci. USA, 68, 1943.
44. Richet, E., Abcarien, P. and Nash, 1986, H.O., Cell, 46, 1011:1021.
45. Rubin, H. and X U K.,1989, "Evidence for the progressive and adaptive nature of spontaneus transformation in the NIH 3T3 cell line", P.N.A.S.-U.S.A., 86, 1860:64 . See Chapter by Nils Ringertz et al. in this book.
46. Staynov, D., 1983, Possible nucleosome arrangements in the higher order structure of chromatin, Int. J. Biol. Neuronal, 5, 3:9.
47. Trask, B. and Van Den Engh, G., Landegent, J., In De Wal, N. J., Van Der Ploeg, M., 1985, Detection of DNA Sequences in Nuclei in Suspension by in Situ Hybridization and Dual Beam Flow Cytometry , Science.
48. Travers, A. and Klug, A., 1987, DNA Wrapping and writhing, Nature, 327, 280:1.

49. Vergani, L., Gavazzo, P., Mascetti, M. and Nicolini, C., Intranuclear DNA distribution and gene expression: in situ hybidrization studies, Hum. Genet, submitted.
50. Vergani, L., Gavazzo, P., Arena, N., Diaspro, A., Facci, P., and Nicolini, C., Cell cycle changes of cytoskeleton related to nuclear DNA: a cytometric analysis of transformed versus fibroblast-like cells, Cell Motility and Cytoskeleton, submitted.
51. Weintraub, N., 1985, Assembly and propagation of repressed and derepressed chromosomal states, (a review) Cell, 43, 705:711.
52. Weintraub, H., Cheng, P., and Conrad, K., 1986, Transcription Driver DNA supercoiling, Cell, 46, 115:122.
53. Wer, H., Shy, S., Wang, J. and Lin, L., 1988, Transcription generates positively and negatively supercoiled in the template, Cell, 53, 433:440.
54. Wolfe, A. P., New approach to chromatin function (review), The New Biologist, 2.
55. Yaneva, I., Zlatanova, J., Srebreva, L., and Jsonev, R., 1990, Interaction ofhistones H1 and H2° with superhelical and linear DNA, FEBS , 263, 225:228.
56. Zacharias, W., O'Connor, T. R. and Larson, J. E., 1988, Methylation of Cytosine in the 5-POsition Alters the Structural and Energetic Properties of the Supercoil-Induced Z-Helix and of B-Z Junctions, Biochemestry, 27, 2970:2978.

PROBLEMS IN UNDERSTANDING THE ORGANIZATION, STRUCTURE AND FUNCTION OF CHROMOSOMES

E. Morton Bradbury

University of California, Davis
Davis, California 95616
Los Alamos National Laboratory
Los Alamos, New Mexico 87545

INTRODUCTION

Despite intensive investigation of mammalian chromosomes, we are still largely ignorant of the basic rules that govern their organization, structure, and functions. This situation results from the current limitations in available technologies to elucidate the structures of such complex biological systems. Whereas the powerful techniques of molecular biology have successfully addressed at high resolution functional problems at the level of nucleic acid sequences, many lower resolution questions concerning the architecture of the cell nucleus, long range order in chromosomes, and higher order chromatin structures remain largely unanswered. Techniques are now emerging that should help to remedy this situation. The use of confocal microscopy with molecular probes will tell us at the level of the light microscope a great deal about the organization of the nucleus and how it changes in different cell types; advanced light sources have the potential to image hydrated biological systems down to 10 nm, and scanning electron tunneling and atomic force microscopies have demonstrated their ability to image molecules though their ability to usefully image biomolecules such as DNA remains to be demonstrated.

CHROMOSOME ORGANIZATION

The human genome contains 6 x 10^9 base pairs (bp) [diploid] or 204 cms of DNA packaged into 46 chromosomes. It is generally believed that chromosomes from all higher organisms each contain a single DNA molecule as has been shown for lower eukaryotes, such as yeasts. Yeast molecular genetics has identified three chromosomal elements essential for replication and equal segregation of chromosomes to daughter cells; i) the ends of the chromosomes or telomeres, ii) the centromere and iii) an origin of DNA replication. In Chinese hamster metaphase chromosomes (Figure 1) the centromere is the constriction at the center of the chromosomes located at the junction of sister chromatids and provides attachment points for the contractile mechanism to separate chromosomes into the daughter cells. Chromosome lengths, locations of centromeres and DNA dye banding patterns are used to identify chromosomes. The origins of the distinctive banding patterns are not well-understood but probably reflect a reproducible pattern of DNA folding induced by the interactions of sequence specific DNA regions with proteins [1]. Chromosomes contain enormous lengths of DNA, e.g., human chromosome 16 contains 3.7 cm DNA molecule packaged into a metaphase chromosome of length 2.5 μm to give a linear

Molecular Basis of Human Cancer
Edited by C. Nicolini, Plenum Press, New York, 1991

DNA packing ratio of 15,000. To account for such high packing ratios earlier models for metaphase chromosomes were based on a series of linear coiled coils. As we shall see a more realistic model to consider is the transverse packing of DNA loops into the thickness of metaphase chromatids.

CHROMOSOMAL PROTEINS AND DNA LOOPS

Isolated chromosomal material consists of long DNA molecules associated with up to twice their weight of chromosomal proteins. Major proteins are the highly conserved group of 5 basic proteins, the histones that are equal in weight to that of DNA. Non-histone proteins are heterogeneous and contain proteins associated with the different chromosome functions, e.g., replication and gene expression as well as those associated with chromosome organization. Some of the proteins associated with long range order in chromosomes have been identified by treating chromosomes with high ionic strengths or with detergents to remove the histones and most other non-histone proteins leaving a small group of proteins tightly bound to the DNA. Electron micrographs of these depleted metaphase chromosomes show quite remarkable structures consisting of a residual protein "scaffold" of the metaphase chromosome surrounded by a halo of DNA [2]. At higher resolution DNA loops can be observed to emerge from and return to the same point on the protein scaffold. (Figure 2) Two major scaffold proteins have been isolated, Sc1 and Sc2. Sc1 has been identified as topoisomerase II [3], which relaxes negatively supercoiled DNA through double stranded DNA cuts, and is essential at metaphase for the separation of sister chromatids to the daughter cells. Recent work from my laboratory shows also that histone kinases and topoisomerase II work in tandem in the process of chromosome condensation [4]. The size range of the DNA loops is 5 kbp to 120 kbp with an average size of about 50 kbp. Thus, the human genome of 3×10^9 bp DNA (haploid) corresponds to 60,000 loops of 50 kbp length that is in the range 50,000-150,000 genes thought to be required to specify a human being. This raises the possibility that each DNA loop defines a genetic unit of one or a small number of linked genes. In *Drosophila* Laemmli has shown that the group of 5 histone genes are located on a 5 kbp loop and that some developmentally regulated genes are single copy genes located on their own DNA loops [5]. The distribution of DNA loops has been shown not to change during development. Thus, questions related to developmental process of gene expression probably involve differential packaging of chromosome loops that determine the availability to factors of those genes required to specify and maintain a particular cell type.

CHROMATIN IS A REPEATING SUBUNIT STRUCTURE

DNA and its associated chromosomal proteins, histones and non-histone proteins, is called chromatin. Digestion of chromatin in nuclei with micrococcal nuclease, which cuts double stranded DNA, gave a ladder of DNA lengths in multiples of a basic unit length of about 190-200 bp [see 6]. This seminal observation showed that chromatin contained a repeating subunit. For most somatic tissues this subunit, the nucleosome contains 195 ± 5 bp DNA the histone octamer $[(H3_2\ H4_2)(H2A,\ H2B)_2]$ and one histone H1. Prolonged, micrococcal nuclease digestion results in well-defined subnucleosome particles; i) the chromatosome with 168 bp DNA, the histone octamer and H1 and ii) the nucleosome core particle with 146 ± 1 bp DNA and the histone octamer [see 6]. The core particle can be obtained in large quantities and has been subjected to extensive structural studies.

Neutron scatter techniques have particular application [7] to core particle structure determination because of the ability, by changing $^1H_2O{:}^2H_2O$ mixtures, to

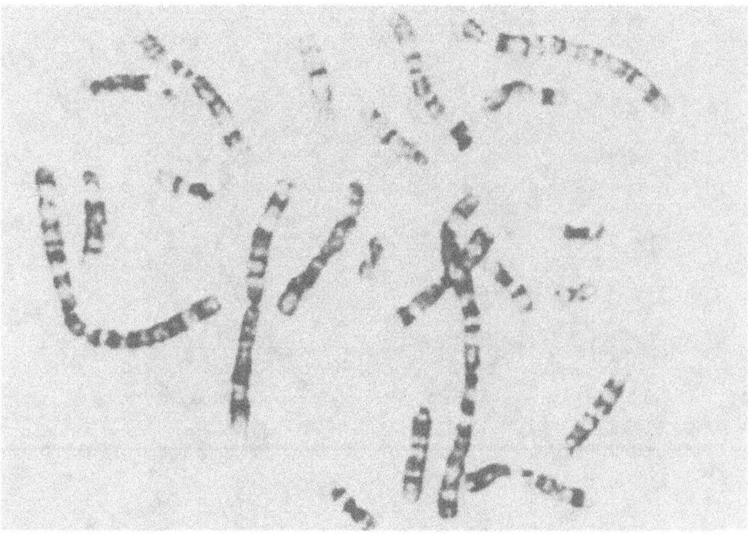

FIGURE 1

Chinese hamster metaphase chromosomes [picture kindly provided by A. Ray, Los Alamos National Laboratory].

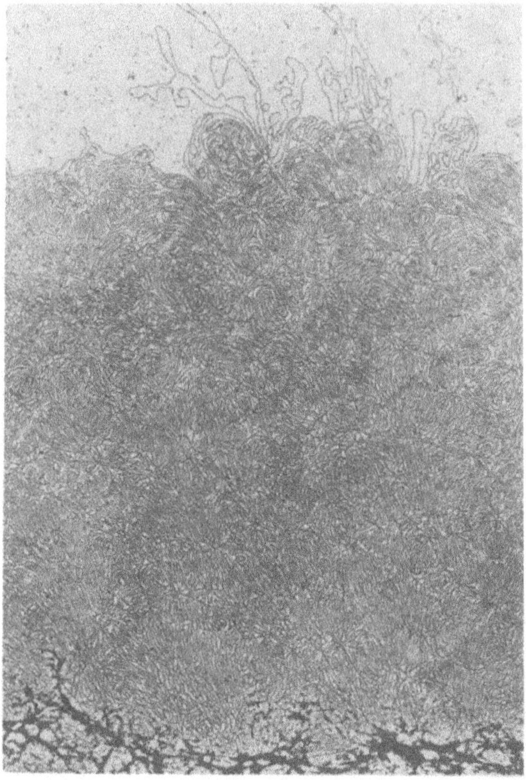

FIGURE 2

Histone depleted metaphase chromosomes showing part of the proteinaceous scaffold and halo of DNA loops [from Paulson J. R. and Laemmli U. K. [2].

"contrast match" either the DNA component (at 65% $^2H_20$) or the protein component (37% $^2H_20$). By neutron contrast matching core particles, the radius of gyration of the DNA was shown to be 5.0 nm while that of the histone octamer was 3.3 nm. These measurements reversed the previously widely held view that histones were complexed on the outside of the DNA. Analysis of the neutron basic scatter functions [8] for the core particle gave the low resolution solution structure of the core particle (Figure 3). The regular structure of the core particle gives crystals that have been solved by x-ray diffraction to 7 Å resolution [9] and 8Å resolution [10]. This structure is identical to that of the solution structure from neutron scatter but at the higher resolution α-helical segments of histones H3 and H4 was observed to interact with the DNA in its shallow groove whereas DNA regulatory proteins recognize and bind their specific sequences in the DNA deep groove. Another significant feature of the 7 Å and 8Å structures is that the DNA does not bend uniformly around the core of the histone octamer but follows a path of less bent segments with tight bends between these segments. This has relevance to the question of nucleosome positioning, i.e., whether some or all nucleosomes are located at precise positions on DNA sequences for functional reasons. DNA footprinting shows that some nucleosomes can be positioned to within a basic pair on a specific DNA sequence, e.g., *Xenopus* 5S rRNA gene DNA can be precisely located on a nucleosome [11]. Analysis of 177 core particle DNA sequences shows that there is a code relating a DNA sequence to its "bendability" [12]. This has relevance not only to the locations of nucleosomes but also to the functions of sequence specific DNA factors controlling gene expression.

CHROMATOSOMES AND NUCLEOSOMES

Extrapolating from the 1.7 turns of DNA of pitch 3.0 nm coiled around the core particle (Figure 3) the chromatosome's 168 bp of DNA is equivalent to 2 turns of DNA. These two turns require the binding of the fifth histone H1. Based on the structure of the core particle, and the conformation of histone H1 a model has been proposed for the chromatosome (Figure 4). A feature of the conformation of histone H1 is that the long flexible "arms" of the molecule have the potential to be involved in long range interactions in chromosome organization. As will be outlined the H1 "arms" undergo a pattern of phosphorylations through the cell-cycle, which would modulate these interactions. The fundamental structural units of chromosomes is the nucleosome. However, to describe the nucleosome beyond the model for the chromatosome requires a knowledge of the paths of the DNA, which links nucleosomes. Those paths are unknown at the present time. This lack of knowledge impedes considerably our understanding of higher order chromatin structures.

HIGHER ORDER CHROMATIN STRUCTURES

Chromatin in low ionic strength aqueous solution is in the form of a 10 nm diameter string of nucleosomes. This form is observed also when chromatin spills out of lysed nuclei. Neutron scatter studies of this 10 nm chromatin fibril give a mass per unit length equivalent to one nucleosome per 10 ± 2 nm, i.e., a DNA packing ratio of between 6 and 7 to 1 [13]. With increase of ionic strength the 10 nm fibril undergoes a transition to the "30 nm" fibrils (Figure 5). Most of the DNA in interphase nuclei and metaphase chromosomes is contained in "30 nm" fibrils. Neutron scatter studies give a diameter for this fibril in solution of 34 nm and a mass/unit length, which is equivalent to 6 to 7 nucleosomes per 11 nm of fibril, i.e., a DNA packing ratio of 40 to 50:1 [13]. Because of the paucity of hard structural data, several models have been proposed for the 34 nm fibril including both one and two start helices. The simplest model that is consistent with available structural data is of a supercoil or solenoid of 6 to 7 radially arranged disc-shaped nucleosomes with a pitch of 11.0 nm and a diameter 34 nm.

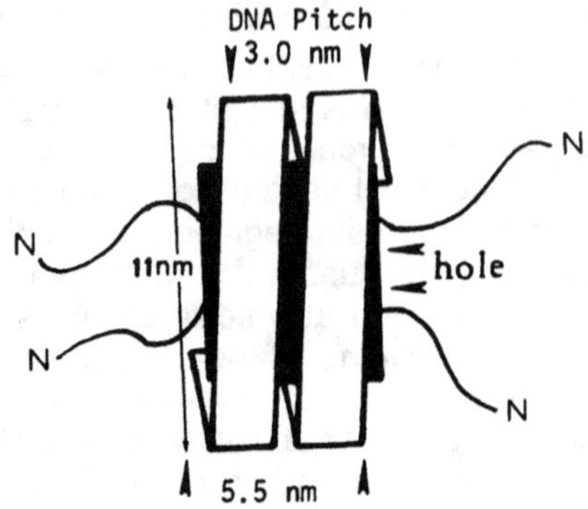

FIGURE 3

Solution structure on the nucleosome core particle from neutron scatter studies. [from Braddock et al. (8)]. The hole is less than 1 nm. The N-terminal regions of histones H2A and H2B are mobile in core particles.

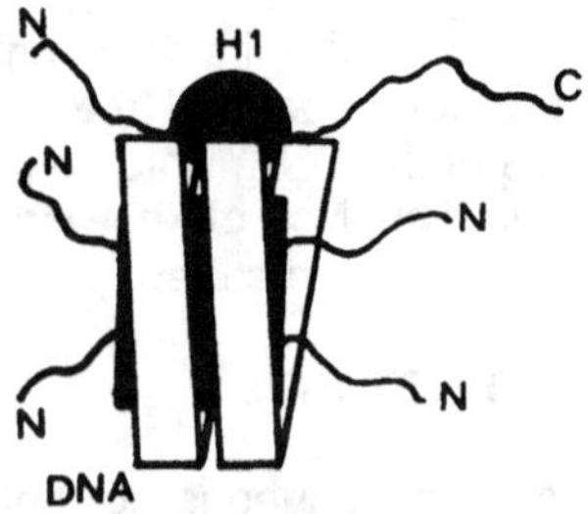

FIGURE 4

Model for the chromatosome with two times of DNA sealed off by the central globular domains of the very lysine rich histone [from Cary et al. (32)].

There is no evidence in the neutron scatter [13] for a hole along the axis of the supercoil of nucleosomes. It is very likely that H1 and the linker DNA are located on the inside of the nucleosome. Because of the paucity of hard structural data on the 34 nm fibril, both one and two start helices have been proposed [see 6]. In the metaphase chromosome the 34 nm filament appears to fold back on itself and coil into a 50-60 nm fibril.

PACKAGING OF CHROMOSOME LOOPS

The EM pictures of histone depleted metaphase chromosomes (Figure 2) show transverse loops of DNA attached to the protein scaffold. The average size of these DNA loops is 50 kbp or 17 μm. With the measured DNA packing ratios, 17 μm of DNA could be packaged into 2.6 μm of the 10 nm fibril or into 0.4 μm of the 34 nm supercoil or solenoid of nucleosomes. Thus, the packaging of a DNA loop into the thickness of a sister chromatid (Figure 1) may require just one more order of chromatin folding. This additional order of folding is probably the 50-60 nm fibril observed in metaphase chromosome.

ARCHITECTURE OF THE CELL NUCLEUS

In a diploid human cell the 2.04 m of DNA is packaged into a cell nucleus about 10^{-5} m in diameter. In general, the size of a cell nucleus is roughly proportional to the amount of DNA it contains. Most cell nuclei have minimum surface to volume ratios, i.e., are spherical. Departures from this shape may occur in specialized cells, e.g., nuclei that are genetically very active probably need to increase their surface area in response to the demand for increased RNA synthesis. Because of the enormous lengths of eukaryotic DNA, it has long been assumed that the cell nucleus is highly organized. So far, however, we have had only glimpses of this order, e.g., telomere organization is maintained through interphase and metaphase chromosomes have the same arrangement in nuclei. There is a protein framework called a nuclear matrix that is thought to play a major role in the organization of the cell nucleus. The nuclear matrix has two major elements; a mesh of protein fibers, called lamins, on the inside surface of the nucleus and an internal protein network. In the current model for DNA organization in the cell nucleus there are specific DNA attachments to both the inner periphery of the nuclear membrane and to the internal protein network. Chromatin takes different forms in different phases of the cell cycle. At metaphase the familiar condensed chromosomes organized by a chromosomal scaffold are observed. Following cell division the chromosomes redistribute throughout the nucleus into condensed chromatin regions (heterochromatin) associated with the inner surface of the nuclear membrane and dispersed chromatin (euchromatin) inside the nucleus. Much of our limited knowledge of the architecture of the cell nucleus comes from electron microscopic studies of cell sections. There is a great need for other microscopic methods that image the hydrated states of nuclei at different stages of the cell cycle and the states of nuclei from different specialized cells.

CHROMOSOME STRUCTURE/FUNCTION RELATIONSHIPS

A major function of the cell division cycle is to replicate faithfully the genome and package it into the set of chromosomes characteristic of the higher organism. Chromosomes undergo major structural transitions through the cell cycle from the functionally active states in S-phase to the condensed inactive states of metaphase chromosome. Major unanswered questions concern the cell cycle controls of these different chromosomal states that are related to chromosome functions. These

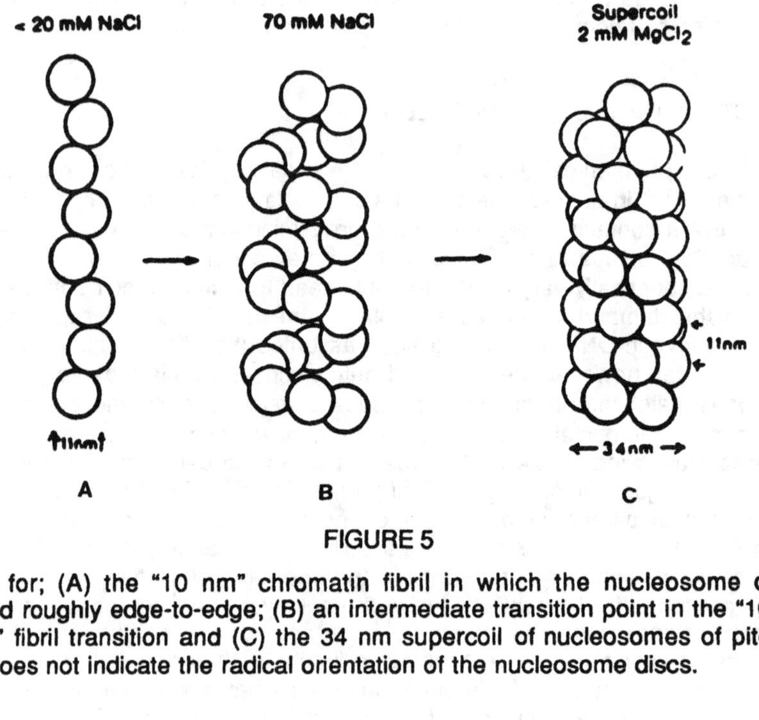

FIGURE 5

Models for; (A) the "10 nm" chromatin fibril in which the nucleosome discs are arranged roughly edge-to-edge; (B) an intermediate transition point in the "10 nm" → "30 nm" fibril transition and (C) the 34 nm supercoil of nucleosomes of pitch 11 nm which does not indicate the radical orientation of the nucleosome discs.

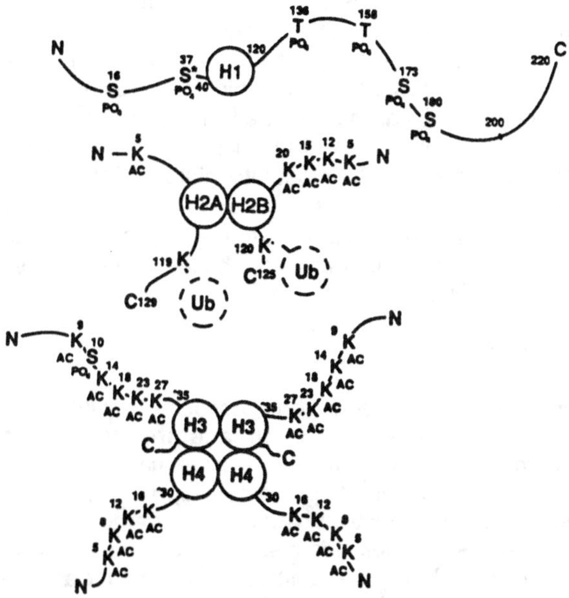

FIGURE 6

Models for the multidomain structures of histones H1, (H2A, H2B) dimer and the ($H3_2$, $H4_2$) tetramer showing the locations of the reversible modifications of acetylation, phosphorylation and ubiquitination.

controls must involve the major structural proteins of the chromosomes, the histones. As outlined above, histones generate the structure of the nucleosome, and the 10 and 30 nm chromatin fibrils. Thus transitions in these different chromatin structural states, in response to functional requirements, must involve modulations of histone DNA interactions and the further interactions of regulating proteins and enzymes that act on DNA.

Histones have well-defined structural domains; rigidly conserved histones H3 and H4 have flexible, basic N-terminal domains extending from apolar globular protein structures; histones H2A and H2B each have a basic, flexible N-terminal domain, a conserved central apolar globular domain and a basic C-terminal "tail" and the fifth histone also has three well-defined domains but with much longer N and C terminal domains. The histone dimer (H2A,H2B) and the histone tetramer ($H3_2$,$H4_2$) are held together by interactions between the apolar globular domains. Outline structures showing the different histone domains are given in figure 6. Shown also are the sites of reversible chemical modifications of histones. There are three major reversible modifications; acetylation of lysines, phosphorylation of serines and threonines and the covalent attachment of a conserved small globular protein, ubiquitin, to the amino group of lysine sidechain. As can be seen, all of these reversible modifications are located in the basic flexible N and C-terminal domains. Such reversible chemical modifications would be expected to modulate histone:DNA interactions by; i) the neutralization of the lysine amino groups in H3 and H4 by acetylation; ii) conversion of serines and threonines is positively charged serine and threonine phosphates and iii) by the introduction of a bulky globular protein, ubiquitin, to lysines sidechain in the C-terminal "tails" of H2A and H2B.

HISTONE ACETYLATION

Histone acetylation has long been associated with all aspects of DNA processing; transcription, DNA replication and the reorganization of DNA in spermiogenesis when histones are replaced by protamines [reviewed in 6] Allfrey first observed the association of histone hyperacetylation with transcriptional activity [14] and has now demonstrated a strict correlation of hyperacetylated H3 and H4 with active genes. Such an association has also been shown for H4 hyperacetylation in Physarum polycephalum [15]. Further antibodies specific for acetylated lysines in the N-terminal of histone H4 bind nucleosomes that are 15-30 fold enriched in active gene sequences [16]. Three distinct patterns of histone acetylations have been observed in the precise nuclear division cycle of physarum polycephalum [17]; i) S-phase acetylation of all the sites in all four core histones that are associated with DNA replication and S-phase gene expression; ii) G2 phase hyperacetylation of only histones H3 and H4 that is associated with G2 phase gene expression and iii) the deacetylation of the four core histones at metaphase presumably to allow for the correct packaging of DNA into the metaphase chromosomes.

Despite their apparent importance in cell biology structure/function relationships of histone acetylation have been elusive. It has been shown that hyperacetylation of the core histones results in a small retardation of core particles on neutron particle gel electrophoresis [18]. Neutron scatter studies have not shown any change in core particle shape on the unwinding of DNA ends of hyperacetylated core particles [19]. Because of these small effects at the core particle level, it was thought that the effects of histone acetylation would have to be sought in fully defined chromatin systems containing arrays of nucleosome particles. This was carried out with a DNA circle of 18 tandem repeats of 207 bp DNA from 5S RNA gene DNA that was shown to contain a nucleosome positioning sequence [20]. From direct measurements of the nucleosome particle linking number change, it was shown that there was a reduction in this parameter with histone hyperacetylation [21] from -1.04 to -0.82 and with the fully acetylated states of histones H3 and H4 only from -1.04 to -0.81 [22]. The implication of this result is that histone acetylation causes the release of negative DNA supercoils from the nucleosome units a constrained DNA loop on domain and thus facilitate the unfolding of a condensed loop for DNA processing.

HISTONE H2A, H2B UBIQUITINATION

Ubiquitination adds a bulky ubiquitin globular protein to H2A and H2B through the ligation of its C-terminal to the amino group of lysine sidechains in the C-terminal tails of these histones. Little or no effect of ubiquitin has been observed on core particle structure leading to the proposal that the ubiquitin moieties would be on the surface of the disc shaped core particles [23]. Such locations would, however, be expected to interfere with the packaging of nucleosomes in the higher order chromatin structure of the 30 nm supercoil of nucleosomes shown in figure 5. Because of this possibility it is of some interest that cell cycle studies of histone ubiquitination show the ubiquitins are removed immediately prior to metaphase and H2A and H2B are reubiquitinated immediately after metaphase [24]. It appears that the perturbing effect of ubiquitination have to be removed to allow the correct packaging of metaphase chromosomes. This suggests that ubiquitin labels an important subcomponent of chromatin possibly containing genes that cannot be permanently repressed but instead must be kept in a potentially active state e.g., stress genes. In this respect it is of interest that active chromatin has been found to be enriched in ubiquitinated H2B [25].

HISTONE PHOSPHORYLATION

Cell cycle dependent phosphorylations of histone H1 and H3 have been observed in both lower and higher eukaryotes. The phosphorylation of H1 has been observed to increase through G2 phase to reach a hyperphosphorylated state at metaphase [26,27]. This phosphorylation parallels chromosome condensation and it has been proposed that this process involves H1 phosphorylation. Cause and effect, however, have yet to be demonstrated. H3 has been shown to be phosphorylated immediately prior to metaphase and is a very late event [28]. It has been shown that the phosphorylation of H1 is preceded by a 15 fold increase in H1 kinase activity and therefore drive the process [26]. In early studies the addition of H1 kinase activity to macroplasmodia of physarum polycephalum three hrs before metaphase advanced mitosis by up to 1 hr [29]. Based on these studies, it was proposed that the eukaryotic cell division cycle was controlled by H1 kinase activity [26,29].

Considerable support to the above proposal concerning the central role of H1 kinase activity in the cell division cycle comes from yeast genetics [30] and also from the identification of the maturation promoting faction (MPF) with H1 kinase [31]. Studies with S. Pombe have identified a central cell cycle control gene cdc2 that codes for a kinase $p34^{cdc2}$ [30,32]. This kinase has been identified in all eukaryotes so far studied. H1 kinase is now known to be a complex formed between p34 and cyclins. p34 is in an inactive state when it is phosphorylated on tyrosine 15 and a threonine [30]. H1 kinase is activated in G2 phase through metaphase by the dephosphorylation of p34. Thus there is an involvement of tyrosine kinases and phosphatases in these cell cycle controls involving H1 kinase. The cyclins are so called because they increase in amounts through the cell cycle and are abruptly degraded immediately following metaphase [31]. This inactivates the H1 kinase releasing p34 to begin another cell cycle.

It is becoming increasingly clear that the cell cycle dependent reversible chemical modifications of histones that must modulate histone/DNA interactions in chromosomes have major roles in the cell cycle control mechanisms. This has been demonstrated convincingly for H1 phosphorylation and the other major modifications of acetylation and ubiquitination are likely to be equally important.

CONCLUSION

We are still a long way from understanding the structure/function relationships of eukaryotic chromosomes despite their central importance to biology. Relevant to their understanding will be the sequence information from the Human Genome Project. Although much interest focuses on the mapping and sequencing of genes, non-coding DNA regions clearly contain sequences involved in the organization and functions of chromosomes. The constancy of banding patterns of individual metaphase chromosomes (figure 1) reflects a highly reproducible pattern of long-range DNA folding, most probably directed by specific DNA-protein interactions and possibly by unusual DNA structure, e.g., bent DNA segments. Superimposed on this long range order is the packaging of the DNA loops, which involves mainly the histones together with other poorly defined structural and regulatory proteins. A major biological question is how the different packaging of loops might determine the states of different gene families, i.e., permanently repressed, potentially active (e.g., heat shock genes) and active genes. DNA control regions of active genes must be accessible to gene regulatory proteins whereas regulatory regions of the permanently repressed genes of a particular cell type may be packaged so that they are inaccessible. Such packaging may also determine the availability of DNA regions to chemical damage. Thus, a knowledge of the organization of chromosomes is essential to an understanding of the central processes of differentiation and development as well as the processes of DNA damage in chromosomes. Such information is also essential to the ultimate understanding of chromosome organization in the cell nucleus. It is to be hoped that the emerging techniques mentioned in the introduction will enable us to address many of these outstanding structural and organizational problems.

ACKNOWLEDGEMENTS

This work is supported by grants from the NIH [PHS GM-26901], the ACS [CD-484F] and the DOE [DE-FG03 88ER60673].

REFERENCES

1. Bickmore, W. A. and Sumner, A. T. (1989) Trends in Genetics 5, 144-148.
2. Paulson, J. R., and Laemmli, U.K. (1977) Cell 12, 817-828.
3. Earnshaw, W. C. and Heck M. M. S. (1985) J. Cell Biol. 100, 1715-1725.
4. Roberge, M., Th'ng, J., Hamaguchi, J. and Bradbury, E. M. (1990) J. Cell Biol. 111, 1753-1762.
5. Gasser, S. M. and Laemmli, U. K. (1987) Trends in Genetics 3, 16-22.
6. van Holde K. E. (1988) Chromatin, Pub. Springer-Verlag, N.Y. (Ed. A. Rich).
7. Bradbury, E. M. and Baldwin, J. P. (1986) in "Supramolecular Structure and Function" (Ed. G. Pifat-Mrzlijak) Springer-Verlag, Berlin, Heidelberg.
8. Braddock, G. W., Baldwin, J. P., and Bradbury, E. M. (1981) Biopolymers 20, 327-343.
9. Richmond, T. J., Finch, J. T., Rushton, B., Rhodes, D., and Klug, A (1984) Nature 311, 532-537.
10. Uberbacher, S. G. and Burnick, G. J. (1989) J. Biomol. Str. Dym. 7, 1-20.
11. Simpson, R. T., Thoma, F., and Brubaker, J. M., (1985) Cell 42, 799-808.
12. Travers, A. A. (1989) Ann. Rev. Biochem. 58, 427-452.
13. Suau, P., Bradbury, E. M. , and Baldwin, J. P. (1979) Eur. J. Biochem. 97, 593-602.

14. Allfrey, V. G. (1980) Cell Biol.: A Comprehensive Treatise 3 (Eds. L. Goldstein and D. M. Prescott) Academic Press Inc.
15. Chabal, S., Mattews, H. R. and Bradbury, E. M (1990) Nature 287, 76-79.
16. Hebbes, T. R., Thorne, A. W. and Crane-Robinson, C (1988) EMBO J. 7, 1395-1402.
17. Matthews, H. R. and Waterborg, J. M. (1985) in "The Enzymology of Post-translational Modifications of Proteins" 2, Academic Press Inc (London) Ltd.
18. Bode, J., Gomez-Lira, N. M. and Schroter, H. (1983) Eur. J. Biochem. 130, 437-445.
19. Imai, B. S., Yau, P., Baldwin, J. P., Ibel, K., May R. P. and Bradbury, E. M. (1986) J. Biol. Chem. 261, 8784-8792.
20. Simpson, R. T., Thoma, F. and Brubaker, J. M. (1985) Cell 42, 709-803.
21. Norton, V. G., Imai, B. S., Yau, P. and Bradbury, E. M. (1989) Cell 57, 449-457.
22. Norton, V. G., Marvin, K. W., Yau, P. and Bradbury, E. M. (1990) J. Biol. Chem. 265, 19848-19842.
23. Kleinschmidt, A. M. and Martinson, H. G. (1981) Nucleic Acids Res. 9, 2423-2431.
24. Muller, R. D., Yasuda, H., Hatch, C. L., Bonner, W. M. and Bradbury, E. M. (1985) J. Biol. Chem, 260, 5147-5153.
25. Nickel, B. E., Allis, D. C. and Davie, J. R. (1989) Biol. Chem. 28, 958-963.
26. Bradbury, E. M., Inglis, R. J. and Matthews, H. R. (1974) Nature 247, 257-261.
27. Mueller, R. G., Yasuda, H. and Bradbury, E. M. (1985) J. Biol. Chem. 260, 5081-5086.
28. Gurley, L. R., Walters, R. and Tobey R. A. (1975) J. Biol. Chem. 250, 3936-3944.
29. Bradbury, E. M. Inglis, R. J. Matthews, H. R. and Langan, T. A. (1974) Nature 249, 553-556.
30. Nurse, P. (1990) Nature 344, 503-507.
31. Lohka, M. J. (1989) J. Cell Sci. 92, 131-135.
32. Pines, J. and Hunter, T. (1990) The New Biologist 2, 389-401.

EXPRESSION OF C-MYC, C-FOS AND C-JUN PROTO-ONCOGENES DURING MUSCLE DIFFERENTIATION IN VITRO

Nils Ringertz, Magnus Rahm, Anna Hultgardh-Nilsson, Pei Jin and Thomas Sejersen

Department of Medical Cell Genetics, Medical Nobel Institute
Karolinska Institutet, Box 60400, 10401 Stockholm

INTRODUCTION

Proto-oncogenes have been shown to code for growth factors, cell surface receptors, membrane proteins, phosphokinases and nuclear proteins. Most of these oncoproteins appear to be part of signalling pathways which elicit or transduce growth regulatory signals from the cell surface to the cell nucleus. Abnormal expression of one or several proto-oncogenes may cause abnormal cell growth. Among the nuclear proteins coded by proto-oncogenes (c-onc) are c-Myc, c-Fos and c-Jun. These proteins have the potential to regulate other genes by interacting directly or indirectly with DNA, thereby influencing gene transcription. The c-fos gene encodes a nuclear phosphoprotein with a half-life of approximately 20 min. Via conserved regions rich in leucine (leucine zipers) the c-Fos protein forms a heterodimer complex with the transcription factor c-Jun/AP-1. The Fos/Jun complex has the ability to bind to regulatory elements of other genes. Both the c-Fos and c-Jun proteins have been found to be families of related proteins which can form hetero- and homodimers with varying affinities for a specific DNA sequence (TGACTCA). The c-myc gene encodes a nuclear phosphoprotein of 58-66kD and similar to the c-Fos and c-Jun proteins it has a short half-life. The c-fos, c-jun, and c-myc genes are activated when quiescent cells are stimulated to proliferate by the addition of growth factors. This event is believed to be important in linking mytogen-induced short term physiological changes to altered patterns of gene expression, cell growth and proliferation.
Immunocytochemical studies indicate that after their synthesis, the c-Fos, c-Jun and c-Myc oncoproteins are rapidly concentrated in the cell nucleus where they are found throughout the dispersed chromatin. The nucleolus does not bind antibodies directed against any of the three oncoproteins. During mytosis the three nuclear oncoproteins are distributed throughout the cytoplasm. At telophase and early G1 phase the oncoproteins reenter the nucleus and are again found throughout the dispersed chromatin. To test if the c-Fos antigen has any tendency to associate with particular interphase chromosomes we analyzed micronucleated cells in which individual chromosomes or groups of chromosomes had formed separate micronuclei. The c-Fos antigen was found in all micronuclei with no apparent affinity for any particular part of the genome[15-16-17]. It should be pointed out that these experiments were made with transfected, c-Fos over-expressing cells, and that c-Fos in normal cells may have a more specific distribution.

In spite of the close association of the c-Myc, c-Fos and c-Jun proteins with the genetic material and the finding that at least c-Fos and c-Jun form complexes that may

Molecular Basis of Human Cancer
Edited by C. Nicolini, Plenum Press, New York, 1991

have a more specific role in gene regulation[23], our understanding of the functional role of oncoproteins is still very incomplete.

Together with some other genes (e.g. actin and ornithine decarboxilase) c-fos, c-jun and c-myc have been characterized as "immediate early response genes" (IER-genes) because they become active within minutes after resting cells have been exposed to a mytogenic stimulus. The speed at which these genes are activated make it less likely that their activation depends on the synthesis of new RNA or protein and more likely that transcription is triggered by a conformational change in the template, translocation of an existing protein from the cytoplasm to the nucleus, and/or to modification of existing proteins, e.g. transcription factors. The upstream regions of the c-myc, c-fos and c-jun genes contain elements which bind regulatory proteins. Some of these elements have been defined by linking upstream promoter regions to reporter genes, and then assying transient expression of the reporter genes in transfected cells. The c-fos promoter region has a serum response element (SRE) capable of binding the phosphorilated form of the serum response factor (SRF), a cyclic AMP response element (CRE), an AP1 site on which the c-Fos protein can bind back to its own gene to form an autoregulatoory loop[22].

Using an established cell line from rat skeletal muscle (L6J1), as well as primary rat myoblast cultures, and early passages of primary smooth muscle cells (SMC) from rat aorta we have tried to analyze the regulation of c-myc, c-fos and c-jun expression during the growth and differentiation of muscle cells. Our work has been focused on early events occurring within the first few hours after quiiescent myoblasts have been stimulated with serum or with growth factors such as platelet-derived growth factor (PDGF) and neurokinin A (NKA).

Previous work has demonstrated that deprivation of mytogenes may bring about precocious myogenic differentiation[12] while stimulation of cell proliferation with fibroblast growth factor (FGF) will delay the onset of differentiation[2-3]. Glucocorticoids and insulin, insulin-like growth factors (IFG I and II) tend to stimulate both proliferation and differentiation of rat myoblasts[5-3]. TGF-b and FGF are able to inhibit differentiation under conditions where no stimulation of cell proliferation occurs[2-11-4]. The effects of growth factors, thus, seems to be complex. However, an important conclusion that can be drawn is that stimulation of proliferation does not necessarily block myogenic differentiation.

PDGF consists of two basic polypeptide chains (A and B) linked by disulfide bonds. The three isoforms AA, AB and BB have been identified and purified from normal tissues. The B-chain is the protein of the C-sis oncogene. The PDGF isoforms bind with different affinities to two distinct subtypes of cell surface receptors. Multiplying myoblasts produce the PDGF A-chain but not the B-chain and they express the PDGF beta-receptor on their cell surface. During myogenic differentiation this receptor is down-regulated. NKA is of interest because this neuropeptide affects not only nerve cells, but also acts as a mytogen for muscle cells. This presentation is a summary of our own work on muscle cells and does not attempt to cover the extensive literature on oncogene function. For recent reviews on this topic the reader is referred to recent articles by Bishop[1], Reddy et al.[18], and Vogt and Bos[23].

FORMATION OF SKELETAL MUSCLE IN VITRO

The main features of in vitro myogenesis are illustrated in Fig.1. When seeded at low cell density, the L6J1 myoblasts will first multiply for several days as mononucleate cells. Then, if the growth medium is appropriate, the cells will exit from the cell cycle and enter a G0 state. After undergoing cell surfaces changes, G0 myoblasts will line up with other G0 myoblasts and fuse to form multinucleated myotubes. This step is followed by changes in gene expression patterns. A large number

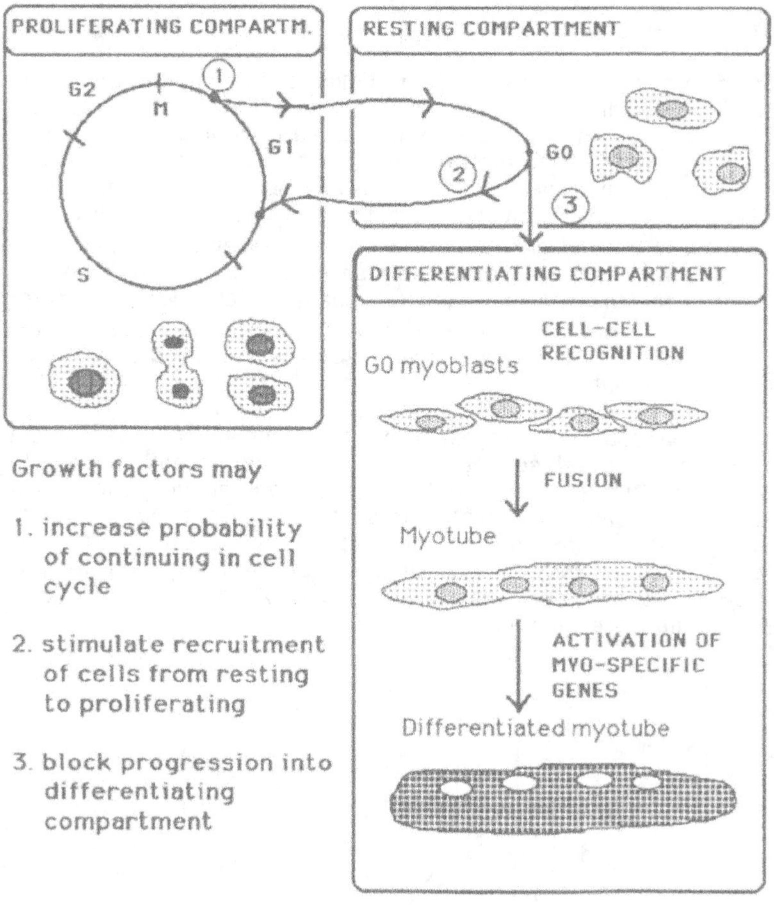

FIG.1 Schematic summary of L6J1 in vitro myogenesis. Growth factors may regulate myogenesis at several points.

of genes coding for muscle specific proteins undergo transcriptional activation. Among the proteins accumulating are the muscle specific isoform of actin, myosin and other contractile proteins which assemble into myofilaments.

Exit of myoblasts from a proliferating into a resting compartment and from there into a differetiating compartment (Fig. 1), appear to be key events in the control of myogenesis. In our work we have tried to examine the role of growth factors and proto-oncogene expression in regulating these transitions.

EFFECTS OF SERUM AND GROWTH FACTOR STIMULATION ON PROTO-ONCOGENE EXPRESSION IN L6 MYOBLAST

Expression of c-fos mRNA could be detected only in poly-A selected RNA in exponentially proliferating L6J1 myoblasts in vitro. The cells do however, contain easily detectable amounts of c-myc and c-jun transcripts. Under normal culture conditions (using 5% FCS) the level of c-myc transcripts decreases when myogenic L6 sublines reach confluency, fuse and form myotubes. Interestingly, differentiation defective L6 sublines which continue to replicate to even higher cell densities, also continue to express the endogenoous c-myc gene[20]. A marked reduction of c-myc and other oncogene transcripts was observed also in C110, a teratoma derived mouse myoblast line, when these cells began to undergo biochemical and morphological differentiation[21]. In more recent work we have focused on the regulation of c-fos, c-jun and c-myc expression in serum-starved myoblasts stimulated to proliferate by the addition of defined growth factors.

A marked introduction of c-fos, c-myc and c-jun mRNA was noted following treatment of serum-deprived myoblasts with recombinant human PDGF-BB homodimers (Fig. 2). c-fos transcripts were present already after 15 min. and maximal levels were found at 30 min. while c-jun and c-myc mRNA peaked at 60 min. Expression of c-fos was transient and decreased rapidly after one hour to return to basal levels. The c-myc and c-jun mRNA levels, on the other hand, remained at an elevated level for several hours. Also NKA treatment of L6J1 myoblasts induced c-fos, c-jun and c-mic mRNA expression.

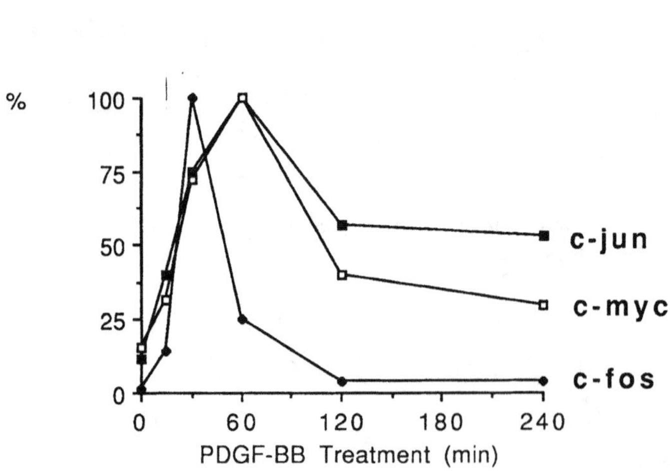

FIG.2 Expression of c-fos, c-juin and c-myc mRNA following treatment with recombinant PDGF-BB.

The c-Fos antigen was induced with kinetics similar to what has been described above for c-fos mRNA. treatment of serum-deprived L6J1 myoblasts with FCS or porcine PDGF-BB induced expression after 30-60 min, as detected using indirect immunofluorescence. Maximal levels were seen after two hours. No induction of c-Fos immunoreactivity could be detected following NKA treatment.

In the presence of low serum concentrations (0,5% FCS), commercially available porcine PDGF-BB transiently stimulated DNA synthesis[8]. However, in cell proliferation assays no effects were observed. In recent experiments[10] recombinant human PDGF-BB has been found to be a much more potent mytogen than biochemically isolated, commercial PDGF-BB of porcine origin. In the presence of human recombinant PDGF-BB, DNA replication and cell proliferation are markedly stimulated, and the cells reach terminal cell densities that are twice as high as those reached by control cultures. Interestingly, the cells fail to differentiate. Both morphological (myotube formation) and biochemical differentiation (creatinine kinase) is effectively blocked by human recombinant PDGF-BB. This effect is reversible. If the cells are shifted to a medium containing 0,5% FCS but no extra growth factors, myotube formation and biochemical differentiation begins.

NKA, although it transiently activates proto-oncogenes c-fos, c-jun and c-myc, fails to stimulate myoblast proliferation and does not significantly affect myogenic differentiation.

ELEVATED C-FOS EXPRESSION REDUCES GROWTH FACTOR REQUIREMENTS AND PARTLY INHIBITS DIFFERENTIATION OF L6 MYOBLASTS

To analyze the relationship between c-fos expression and L6J1 in vitro myogenic proliferation and differentiation, L6J1 rat skeletal myoblasts were transfected with expression vectors containing the murine c-fos gene. L6J1 cells with elevated expression of transfected c-fos reached higher terminal cell densities and had faster growth rates than had control myoblasts trasfected with constructs containing only a neomycin resistance gene (neo-transfected cells) but no proto-oncogene. The morphology of c-fos transfected L6J1 myoblasts was not visibly altered from that of control myoblasts. Formation of myotubes was, however, significantly reduced in c-fos transfected cultures compared with neo-transfected controls. Myotube formation and expression of the myogenic markers alpha-actin and myosin heavy chain were reduced in subclones expressing high level of c-fos, but not in subclones with lower levels of c-fos expression[14].

CADMIUM INDUCES C-MYC AND C-JUN BUT NOT C-FOS IN L6 MYOBLASTS

In transfection experiments with constructs containing the c-fos gene under a metallothionein promoter, it was noted that neo-transfected control cells and normal myoblasts responded to heavy metals by transcriptional activation of c-jun and c-myc. Cadmium alone was found to induce a transient accumulation of c-jun and c-myc mRNA with maximum expression at 2-4 hours[10]. At the same time, the level of c-fos transcripts remained below the detection level. Both the c-fos and the c-jun genes could, however be induced by treating the myoblasts with insulin. Cadmium induction of c-jun and c-myc mRNA occurred in a concentration-dependent manner with a maximum stimulation at 5-10 mM. In the presence of cycloheximide, c-jun and c-myc genes were superinduced by the addition of cadmium. Under these conditions there was also a marked increase in c-fos transcripts. The fact that the c-fos gene can be induced by cadmium only if protein synthesis is blocked by cycloheximide and the observation than the other genes are superinduced by cycloheximide suggests that they are regulated by labile repressors. Furthermore, the observation that the three proto-oncogenes differ in their response to cadmium and insulin suggests that these agents may exert their activity via different response elements and/or different signalling pathways. (Fig.3).

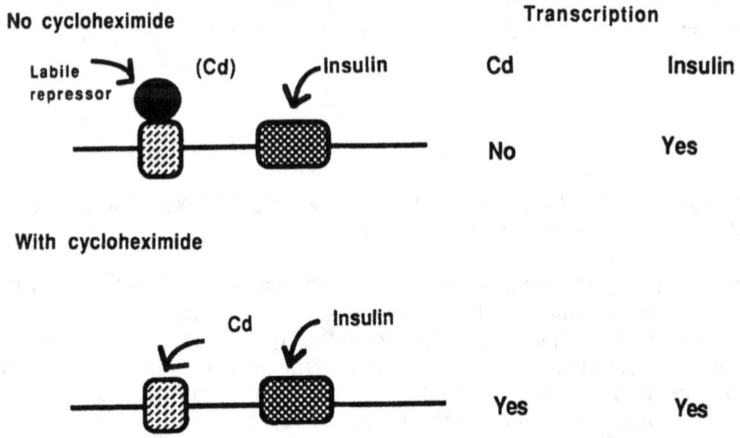

FIG.3 Hypothetical scheme for explaining the ability of cadmium ions to induce c-fos transcription in the presence of cycloheximide. Cadmium and insuline may regulate c-fos expression via two different regulatory elements in the c-fos promoter. The cadmium sensitive element may differ from the insulin sensitive element in being regulated by a labile repressor which cannot be synthesized in the presence of cycloheximide.

Induction of c-myc and c-jun by cadmium, and of c-fos by a combination of cadmium and cycloheximide, could be abolished by blocking transcription with actinomycin D. the cadmium-induced increase in c-jun and c-myc mRNA was enhanced in myoblasts stably transfected with a mouse c-fos gene under a metallothionein promoter. Our present data show that cadmium has the potential to deregulate the expression of several the expression of several important proto-oncogenes.

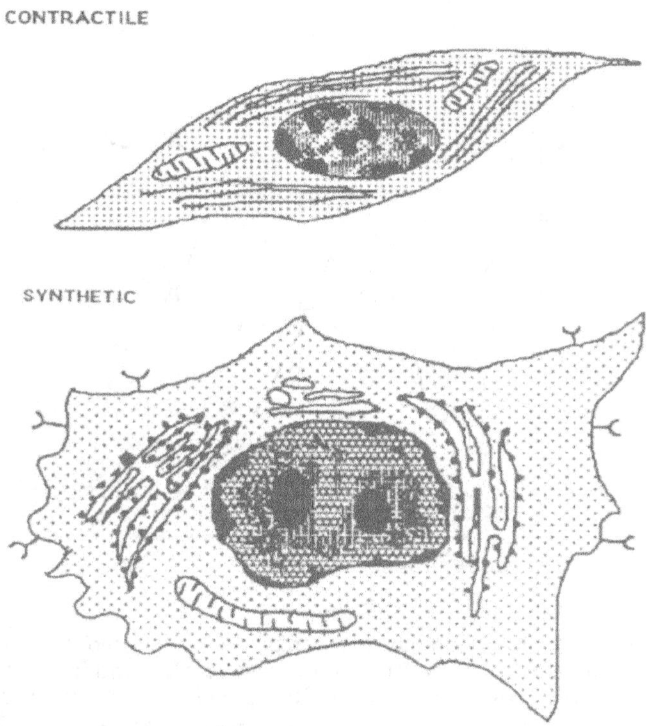

FIG.4 Schematic illustration of the two main phenotypes shown by smooth muscle cells from rat aorta

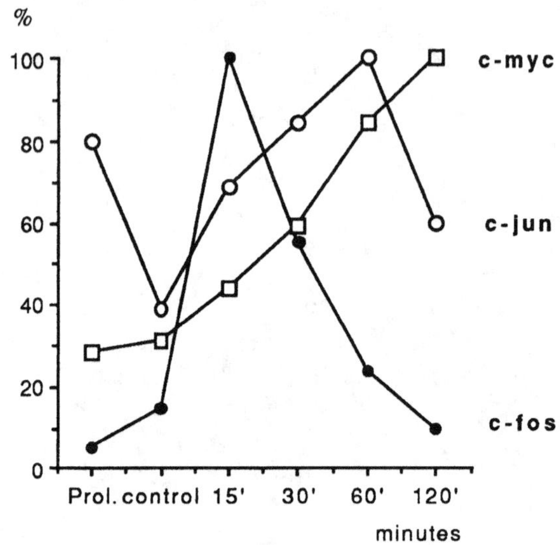

FIG.5 Induction of proto-oncogene expression in adult smooth muscle cells by neurokinin A. Proliferating cells ("prol") express c-myc and c-jun but not c-fos; Serum starved cells ("control") show down-regulation of c-myc and c-jun. After stimulation with neurokinin A there is a sharp increase in c-fos transcripts (from[7]).

REGULATION OF C-MYC, C-FOS AND C-JUN IN SMOOTH MUSCLE CELLS (SMC)

Smooth muscle cells (SMC) are found in the contractile portion of vessels, the digestive tract, the urinary and genital ducts and in the skin. Although the main function of SMC is to contract they may also, after changing from a contractile to a synthetic phenotype, begin to proliferate and to take part in tissue regeneration. The phenotypic change which accompanies the initiation of cell proliferation is usually referred to as a "dedifferentiation" phenomenon, but it may in fact be reversible and might alternatively then be described as a case of "phenotypic modulation". However, for practical purposes the phenotypic change will be referred to as dedifferentiation (Fig.4).

Dedifferentiation of SMC from a contractile to a synthetic phenotype is an important feature in pathological disorders involving prolifertion of SMC such as is the case in atherosclerotis and leimyosarcoma. Dedifferentiation also occurs when primary SMC from adult animals are cultured in vitro. By contrast SMC of the newborn, show a synthetic phenotype already in the intact tissue. When these cells are explanted in cell cultures they are already in a synthetic state and will start proliferating without undergoing a dedifferentiation process in vitro. SMC isolated from the aorta of newborn and adult rats, therefore, constitute an interesting experimental model for studies of mechanisms regulating smooth muscle cell differentiation.

In vitro, newborn and adult SMC differ from each othe in several aspects. Some of the differences are quite unexpected. When grown in standard, serum-containing medium, SMC from adult animals proliferate at a higher rate and grow to a higher finaldensity than do SMC from newborn animals. Adult SMC also demonstrate an increased rate of DNA synthesis in response to platelet-derived growth factor AA and BB in spite of the fact that SMC from young and adult animals appear to have similar numbers of surface receptors for platelet-derived growth factor. The regulation of proto-oncogene transcription also appear to be different in newborn and adult SMC. Both cell types express comparable levels of c-myc and SMC-specific alpha-actin mRNA (a differentiation marker in SMC) when grown in the presence of 10% serum. However, transfer of the cells to serum-free medium results in a marked increase in c-myc and alpha-actin transcripts in newborn SMC, whereas the expression in adult SMC shows a slight decrease. Serum-stimulation leads to a transient increase in c-fos and c-jun mRNA expression in both young and adult cells.
We have earlier demostrated[13] that the sensory neuropeptide Neurokinin A (NKA) stimulates growth of cultured adult SMC, suggesting that sensory neurons participate in the growth regulation of surrounding tissues. NKA, at mitogenic concentrations, was also found to stimulate formation of inositol phosphates[6], to induce cytoplasmic alkalinization and to induce expression of the c-fos, c-jun and c-myc genes (Fig. 5).
Pretreatment of the cells with the calcium antagonist nisoldipine or the Na^+/H^+ exchange inhibitor amiloride inhibited NKA induced proto-oncogene expression as well as DNA synthesis, indicating that both the inositol phosphate-mediated release of calcium ions and diacylglycerol-mediated Na^+/H^+ exchange are involved in the signals evoked by NKA.

CONCLUDING REMARKS

1. Treatment of serum-deprived L6J1 myoblasts with recombinant human PDGF-BB or with the neuropeptide NKA induces increased levels of c-fos, c-jun and c-myc mRNA. With both growth factors the increase in c-fos is transient and lasts only for about an hour while the abundancy of c-jun and c-myc transcripts remains at an elevated level for several hours. In the presence of low serum concentrations, or if serum is replaced by serum albumin, recombinant human PDGF-BB also induces DNA synthesis and cell proliferation, in L6J1 rat myoblasts. Furthermore, the addition of PDGF causes a complete, but reversible, block in morphological and biochemical

muscle differentiation. Our results show that PDGF is an important factor in the regulation of in vitro myogenesis.

2. Increased expression of c-jun and c-myc mRNA can be induced by cadmium ions. If protein synthesis is inhibited by cycloheximide when the cells are treated with cadmium there is also a dramatic increase in c-fos transcripts and superinduction of c-myc and c-jun. These results suggest that labile repressors may be important in the regulation of these three proto-oncogenes. The mechanism by which cadmium activates proto-oncogenes is not known. There is evidence that cadmium induces changes in the metabolism of inositol phosphates and a release of bound cytoplasmic calcium. Cadmium, therefore, induces some of the cytoplasmic secondary messages seen when quiescent cells are stimulated with growth factors. While it seems likely that cadmium deregulates proto-oncogene expression by interfering with normal signal transduction pathways it is not possible to define at which level this effect occurs. Cadmium may for instance compete with calcium for binding to calcium channels or to cytoplasmic calcium-binding proteins or with zinc for nuclear proteins containing DNA-binding zinc-fingers. Furthermore, it cannot be excluded that in addition to such generalized effects, the promoter regions may contain metal sensitive response elements (Fig.3).

The observation that cadmium ions deregulate proto-oncogenes is interesting for several reasons. Firstly, it can help explain the carcinogenic and toxic effects of a metal which presents serious problems as an environmental pollutant. Secondly, cadmium can be used as an experimental tool to analyze signalling pathways regulating proto-oncogene expression. Thirdly, it is a warning to cell geneticists using cadmium to induce transfected genes. Cadmium, in addition to inducing genes in constructs containing metallothionein promoters, also causes deregulation of cellular genes.

3. In c-fos-transfected L6J1 myoblasts with constitutively elevated c-fos expression, the response to mitogenic stimulation is facilitated and terminal differentiation is partly inhibited.

4. Smooth muscle cells (SMC) isolated from adult rats proliferate at a higher rate than SMC isolated from newborn rats and differ in their proto-oncogene response to growth factor stimulation. These differences are interesting in relation to aging in general and atherosclerosis in particular. SMC proliferation is an early event in the development of atherosclerotic lesions in the arterial wall.

5. Neurokinin A (NKA) stimulates formation of inositol phosphates, induces cytoplasmic alkalinization, activates transcription of the c-fos, c-jun and c-myc genes and induces DNA synthesis in adult SMC. NKA appears to be a more potent mitogen for smooth muscle cells than for L6J1 skeletal muscle cells. In the latter case NKA induced proto-oncogene transcription but failed to stimulate DNA replication and cell proliferation.

6. The finding that cadmium ions deregulate proto-oncogene expression and that drugs acting on ion channels can block growth factor induced transcription from these genes, suggest that ionic signals are important in signalling pathways regulating c-fos, c-jun and c-myc transcription. Other lines of evidence indicate that protein phosphorylation and translocation of proteins from the cytoplasm to the nucleus are important factors in the regulation of gene expression (Fig. 6). Clearly further work is needed to elucidate how these phenomena are related to the transduction of signals regulating proto-oncogenes and other immediate early response genes triggered by mytogenic stimulation.

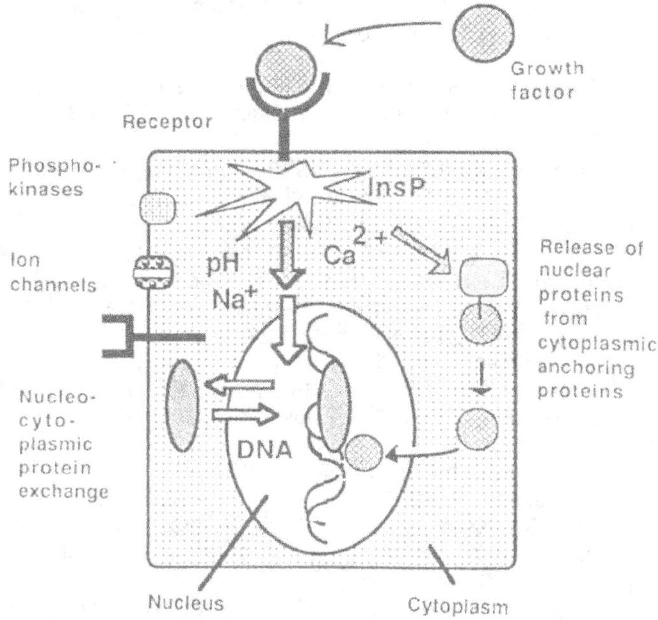

FIG. 6 Schematic illustration of signalling pathways that may be involved in triggering proto-oncogene expression after growth factor stimulation. The nucleus may (1) respond directly to ionic signals, there may be a release of nuclear proteins from cytoplasmatic anchoring proteins , (3) alterations in nucleocytoplasmatic protein trafficking and/or (4) modification of transcription factors already bound to DNA.

REFERENCES

1. Bishop, J.M. (1987). The molecular genetics of cancer,. Science 235, 305-311.
2. Clegg, C.H., Linkhart, T.A., Olwin, B.B. and Hauschka, S.D., 1987,. Growth factor control of skeletal muscle differentiation: commitment to terminal differentiation occurs in G1 phase and is repressed by fibroblast growth factor., J. Cell Biology, 105, 949-956.
3. Ewton, D.Z. and Florini, J.R., 1981,. Effects of the somatomedins and insulin on myoblast differentiation in vitro, Develop. Biol., 86, 31-39.
4. Florini, J.R. and Magri, K.A.,1989, Effects of growth factors on myogenic differentiation, Am. J. Physiol. ,Cell Physiol. 25, C701-C711.
5. Guerriero, V. Jr., and Florini, J.R., 1980, Dexamethasone effects on myoblasts proliferation and differentiation, Endocrinology, 106, 1198-1202.
6. Hultgardh-Nillson, A., Nillson, J., Jonzon, B., and Dalsgaard, C.J., 1988, Coupling between inositol phosphate formation and DNA synthesis in smooth muscle cells stimulated with neurokinin A., J. Cell Physiol., 137, 141-145.
7. Hultgardh-Nillson, A., Larsson, S., Jin, P., Sejersen, T., and Ringertz, N.R., 1990,. Neurokinin A induces expression of the c-fos, c-jun and c-myc genes in rat smooth muscle cells. (submitted for publication).
8. Jin, P., Rahm, M., Claesson-Welsh, L., Heldin, C-H, and Sejersen, T., 1990, Expression of PDGF A-chain and beta-receptor genes during rat myoblast differentiation, J. Cell Biol ,110, 1665-1672.

9. Jin, P., Sejersen, T., and Ringertz, N.R.. Recombin,ant PDGF-BB stimulates growth and inhibits differentiation of rat L6 myoblasts. (submitted for publication).
10. Jin, P., and Ringertz, N.R., 1990, Cadmium induces transcription of proto-oncogenes c-jun and c-myc in rat L6 myoblasts, J. Biol. Chem. (accepted for publication).
11. Lathrop, B., Olson, E., and Glaser, L., 1985, Control by fibroblast growth factor of differentiation in the BC3H1 muscle cell line, J. Cell Biol., 100, 1540-1547.
12. Nadal-Ginard, B., 1978, Commitment, fusion and biochemical differentiation of a myogenic cell line in the absence of DNA synthesis, Cell 15, 855-864.
13. Nilsson, J., Sejersen, T., Hultgardh-Nilsson, A., and Dalsgaard, C-J ,1986, DNA synthesis induced by the neuropeptide substance K correlates to the level of myc-gene transcripts, Biochem. Biophys. Res. Comm., 137, 167-174.
14. Rahm, M., Jin, P., Sumegi, J., and Sejersen, T., 1989, Elevated c-fos expression inhibits differentiation of L6 rat myoblasts, J. Cell Physiol., 139, 237-244.
15. Rahm, M., and sejersen, T., (1990). c-fos reduces growth factor requirements for mytogenic stimulation of L6 rat myoblasts, J. Cell Physiol., 142, 352-358.
16. Rahm, M., and Hultgardh-Nilsson, A., 1990, Neurokinin A induces c-fos expression in L6 rat myoblasts (manuscript in preparation).
17. Rahm, M., Hultgardh-Nilsson, A., Jiang, W-Q, Sejersen, T., and Ringertz, N.R., 1990, Intracellular distribution of the c-fos antigen during the cell cycle, J. Cell Physiol., 143, 475-482.
18. Reddy, E.P., Skalka, A.M., and Curran, T., 1988, in "The oncogene handbook", Elsevier Science Publishers, New York.
19. Ringertz, N.R., Krondal, U., and Coleman, J.R., 1978, Reconstitution of cells by fusion of cell fragments, Exptl. Cell. Res., 113, 233-246.
20. Sejersen, T., Sumegi, J., and Ringertz, N.R., 1985, Density-dependent arrest of DNA replication is accompanied by decreased levels of c-myc RNA in myogenic but not in differentiation-defective myoblasts, J. Cell Physiol., 125, 465-470.
21. Sejersen, T., Betsholtz, T., Sjolund, M., Heldin, C-H, Westermark, B., and Thyberg, J., 1986, Rat skeletal myoblasts and arterial smooth muscle cells express the gene for the A-chain but not the gene for the B-chain (c-sys) of PDGF and produce a PDGF-like protein, Proc. Natl. Acad. Sci., 83, 6844-6848.
22. Shaw, P.E. Hipskind, R.A. Schroter, H., and Nordheim, A., 1989, Transcriptional regulation of proto-oncogene c-fos in "Nucleic Acid and Molecular Biology", Vol. 3, Ed. Eckstein F. and Lilley DMJ Springer-Verlag, Berlin.
23. Vogt, P.K. and Bos, T.J.,1989, in "The oncogene jun and nuclear signalling", TIBS 14, 172-175.

THE EXPRESSION AND REGULATION OF PP90^{RSK}, A HIGHLY CONSERVED MITOGEN-RESPONSIVE SERINE-SPECIFIC PROTEIN KINASE

R.L. Erikson*, David A. Alcorta, Eleanor Erikson§, Laurel J. Sweet°* and Terry A. Vik*

*Dep. of Cellular and Developmental Biology, Harvard Univ.,16, Divinity Av., Camb., Mass., 02138; §Dep. of Pharmacology, Univ. of Colo. Sch. Medicine, 4200 East Ninth Av., Denver, Colo. 80262; °Calif. Biotech., Inc., 2450 Bayshore Parkway, Mountain View, Calif. 94043

INTRODUCTION

Reversible protein phosphorylation is believed to play a major role in the response of cells to mitogens. Although in many cases the initial event involves the activation of tyrosine-specific protein kinases, there is also a rapid activation of serine/threonine protein kinases. Indeed, in vivo tyrosine phosphorylation of proteins is rarely seen in the absence of additional modification by serine/threonine phosphorylation. We have been interested for the past several years in the molecular events that lead to activation of serine/threonine kinases.

Several serine/threonine-specific protein kinases have ben implicated as potential substrates for tyrosine-specific protein kinases, including MAP kinase[1], casein kinase II[2], cdc2/histone H1 kinase[3], the oncogene product Raf-1[4] and S6 kinases[5-7]. Under appropriate conditions MAP kinase and cdc2/histone H1 are phosphorylated on tyrosine and the former enzyme is activated, whereas the latter is inactivated, by this modification. In each case the enzymes are also phosphorylated on threonine apparently to an extent equal to that of tyrosine. Both sites may need to be modified on MAP kinase to achieve a funtional change[8]. The Raf-kinase has recently been shown to be activated by PDGF receptor phosphorylation in vitro[9] and is perhaps a substrate for pp60$^{v\text{-}src}$ in vivo. Casein kinase II is activated by mitogen stimulation, but there is no evidence that it is phosphorylated directly by growth factor receptors. With the exception of genetic evidence on the function of cdc2/histone H1 kinase in the cell cycle of yeast, the significance of any phosphorylation mediated by these serine/threonine-specific kinases remains to be determined. The function of S6 phosphorylation is also not clarified, but its phosphorylation is associated with transition from G0 to G1 in the cell cycle and may enhance the protein synthetic capacity of ribosomes.

We have obtained cDNA clones for a major serine-specific enzyme originally identified as S6KII[10]. In this talk, I will review our recent progress on our studies of the product of this gene, denoted rsk (for ribosomal S6 kinase), pp90. Readers should note that most of the data presented here has been previously published in other communications from this laboratory[11-13].

RESULTS AND DISCUSSION

An S6 kinase of M_r 92,000, denoted S6KII, has been purified from unfertilized Xenopus eggs[10]. The activated form of this enzyme can be inactivated by dephosphorylation and partially reactivated by MAP kinase in vitro[14]. Partial peptide sequences were used to identify two Xenopus cDNA clones corresponding to mRNA for S6KII. These clones predicted proteins of M_r 83,000 and 71,000, that were designed S6KIIa and S6KIIb, respectively[15].

Activation of S6KIIa in insect cells coinfected with baculovirus expressing $pp60^{v\text{-}src}$

The need for sufficient $pp90^{rsk}$ protein for studies on its activation prompted us to use the baculovirus vector system to produce recombinant proteins. Recombinant baculoviruses expressing Xenopus rsk-a and rsk-b protein, which is truncated at the COOH terminus by some 104 amino acids compared to rsk-a, were used to infect Sf9 cells. As Fig. 1A shows, virus-infected cells, when labeled with [^{35}S]methionine at 36-40 hours after infection, had reduced synthesis of most cellular proteins (lanes 2-4) when compared to uninfected cells. Infections by the recombinant rsk-a or -b producing viruses resulted in unique bands at M_r 83,000 and M_r 71,000 (lanes 2 and 3, respectively) when compared to the wild-type infected cells (lane 4). Immunoprecipitation demonstrated that these proteins were recognized by anti-rsk serum (Fig. 1B, lanes 2 and 3). These molecular weights are similar to those of the products of in vitro transcription and translation of the rsk-a and rsk-b cDNAs[16].

The functional activity of the kinases produced in Sf9 cells was noted initially by their ability to autophosphorylate. The signal seen by labeling with [g-^{32}P]ATP was much less intense than expected, since sufficient quantities of rsk proteins are produced by the cells to result in detectable Coomassie stainable protein upon immunoprecipitation. The phosphorylation signal seen in rsk-b was less than that seen in rsk-a, perhaps because of a loss of an autophosphorylation site in the truncated protein or lower specific activity of the truncated form. Moreover, initial attempts to show specific phosphorylation of the S6 protein in 40S ribosomes were unsuccessful.

We attempted to mimic the pathway of activation reported for RSV-transformed chicken embryo fibroblasts by simultaneously infecting Sf9 cells with recombinant baculoviruses containing the cDNA for expression of $pp60^{v\text{-}src}$ and either rsk-a or rsk-b. Autophosphorylation of immunoprecipitates of cell lysates from uninfected Sf9 cells is shown in Fig. 2A. The background activity in uninfected and wild-type infected cells was quite low (lanes 1 and 7). The rsk-b protein (lanes 5 and 6) showed significantly less activity than rsk-a with or without v-src expression. Any activity detected is apparent only on prolonged exposure of the autoradiograph. Coproduction of v-src and rsk-a resulted in greatly enhanced autophosphorylation of rsk-a compared to rsk-a alone (lane 4) and resulted in heterogeneous migration of rsk-a. The most slowly migrating form of phosphorylated rsk-a had an approximate Mr 90,000, similar to the M_r of purified Xenopus S6KII. Coinfection with v-src and rsk-b (lane 5) did not result in a detectable change in the migration of rsk-b. When immunoprecipitates of these lysates were used to phosphorylate 40S ribosomes (Fig. 2B), only the doubty infected v-src and rsk-a sample (lane 3) showed significant phosphorylation of the S6 protein. Prolonged exposure of the gel revealed no phosphorylation of S6 above background in the remaining lanes, sugesting at least a 100-fold increase in the activity od rsk-a as the result of $pp60^{v\text{-}src}$ expression.

In order to determine whether other src gene products would have a similar effect, baculoviruses expressing $pp60^{c\text{-}src}$ or $pp60^{NY315}$ were co-produced with rsk-a in Sf9 cells (Fig. 2C). Again, immunoprecipitates of lysates from double-infected cells were used to phosphorylate 40S ribosomes. Lane 1 shows the activity in v-src and rsk-a-infected cells. The activity in c-src- and rsk-a-infected cells (lane 2) shows significantly less phosphorylation of S6 but was above the level of activity in

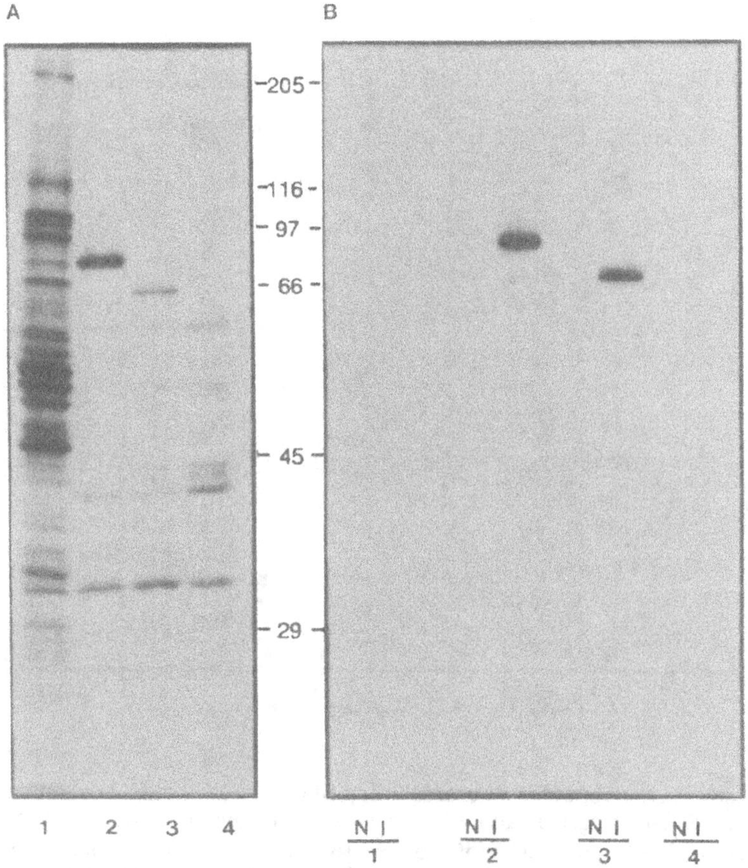

FIG. 1. [^{35}S]methionine labeling and immunoprecipitation of rsk products from insect cells. Recombinant baculoviruses encoding rsk-a or rsk-b were used to infect Sf9 cells. Labeled lysates were resolved directly by SDS/PAGE or immunoprecipitated with either preimmune or immune serum, then resolved. A.) Direct analysis of labeled protein from uninfected (lane 1), rsk-a infected (lane 2), rsk-b infected (lane 3), or wild-type infected Sf9 cells (lane 4). B.) Immunoprecipitation of labeled protein from uninfected (lane 1), rsk-a infected (lane 2), rsk-b infected (lane 3), or wild-type infected Sf9 cells (lane 4). N, Pre-immune serum, I, anti-rsk serum. Molecular weight standards in kDa are indicated.

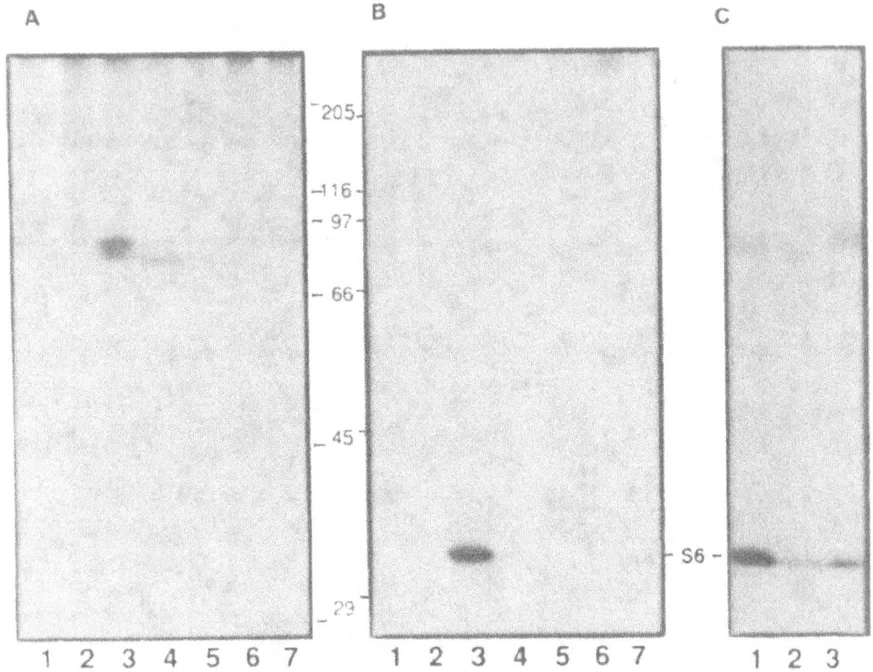

FIG. 2. Activation of pp90^{rsk}. Lysates from Sf9 cells 28 hours post infection were immunoprecipitated, washed, and allowed to autophosphorylate or phosphorylate 40S ribosomes in the presence of [g-^{32}P]ATP. A.) Autophosphorylation in immunoprecipitates from uninfected (lane 1), v-src infected (lane 2), v-src + rsk-a infected (lane 3), rsk-a infected (lane 4), v-src + rsk-b infected (lane 5), rsk-b infected (lane 6), or wild-type infected (lane 7) sF9 cells. B.) Phosphorylation of 40S ribosomal subunits by immunoprecipitates. Lanes are identical to those in panel A. C.) Phosphorylation of 40S ribosomal subunits by immunoprecipitates from lysates of Sf9 cells which were v-src + rsk-a infected (lane 1), c-src + srk-a infected (lane 2), or NY315 + rsk-a infected (lane 3). Molecular weight standards in kDa and S6 migration are indicated.

rsk-a alone infected cells (data not shown). The activity in NY315 and rsk-a-infected cells (lane 3) was intermediate but reproducibly lower than in v-src and rsk-a cells. The protein levels of the various src-gene products and $pp90^{rsk}$ were equivalent upon immunoblot analysis (data not shown). More rsk protein was found in the higher molecular weight forms in the v-src and NY315 than in the c-src double-infected cells.

Although the v-src and rsk-a doubly infected cells had S6 kinase activity, it appeared that most of the rsk-a protein produced in these cells comigrated with the inactive protein, as judged by Coomassie blue staining. A purification scheme was devised to separate the active protein from the unactivated forms (Fig.3A and 3B). All forms of protein produced by recombinant virus bound to S-sepharose® fast flow were eluted with a salt gradient. Representative fractions from the column were resolved by SDS/PAGE and immunoblotted with anti-rsk serum. As seen in Fig. 3A, the slower migrating forms (fraction 27) were eluted earlier than those forms which migrate at the predicted molecular weight (fraction 41). More significantly, Fig. 3B demonstrates that the majority of the S6 kinase activity resided in the early fractions containing the slowest migrating forms of $pp90^{rsk}$, suggesting that only the most slowly migrating form of the kinase makes a significant contribution to S6 phosphorylation. Other experiments have shown the slowly migrating forms are more highly phosphorylated.

Identification of a homolog of S6KII in cultured cells

Several laboratories have reported the purification of proteins with S6 kinase activity from different tissue sources. In addition to the well-studied S6KII, enzymes of approximate M_r's of 65,000-70,000 have been identified in 3T3 fibroblasts, chicken embryo fibroblasts (CEF), bovine liver, regenerating rat liver, and chicken embryos[5-7]. Until recently, the relationship between these proteins remained unclear; however, the recent cloning of chicken and mouse homologs of the Xenopus ribosomal S6KII gene predicts the presence of an 84,000-M_rS6 kinase in mouse and chicken[16].

Polyclonal antibodies raised against purified Xenopus S6KII have been shown to immunoprecipitate an S6 kinase activity from CEF[17]; however, this antiserum was unable to detect a specific protein that could have accounted for the activity obsrved. A polyclonal antiserum against recombinant rsk product produced in bacteria was also raised in rabbits (denoted 125). Cell-free lysates from CEF serum-stimulated for 1 hour were analyzed by immunoblotting with antiserum 125, and the results presented in Figure 4 indicate the presence of a homolog of Xenopus S6KII in CEF. To determine whether the 90,000-M_r protein recognized by antiserum 125 was an S6 kinase, immunocomplexes formed between CEF cell lysates and 125 antisrum were assayed for S6 kinase activity and compared to assays of S6 kinase activity measured in the same cell lysates without immunoprecipitation. As shown previously, serum-stimulation of CEF resulted in an increase in phosphorylation of S6 in 40S ribosomal subunits exogenously added to cell-free lysates. Similar cell lysate preparations were immunoprecipitated with 125 antiserum, followed by assay of S6 phosphorylation. The results demonstrated the presence of an immune-specific S6 kinase activity in CEF and, further, that this activity was greater in immunocomplexes formed from serum-stimulated CEF lysates (data not shown). When immunocomplexes formed between 125 antisrum and cell lysates of control or serum-stimulated CEF were incubated with [g-^{32}P]ATP in the absence of expogenous substrate, a 90,000 M_r protein was radiolabeled.

As mentioned above, the activation of S6 kinase in mitogen-stimulated cells involves phosphorylation of the enzyme. The effect of mitogen-stimulation on the phosphorylation of pp90 was examined in CEF biosynthetically labeled with $H_3{}^{32}PO_4$.

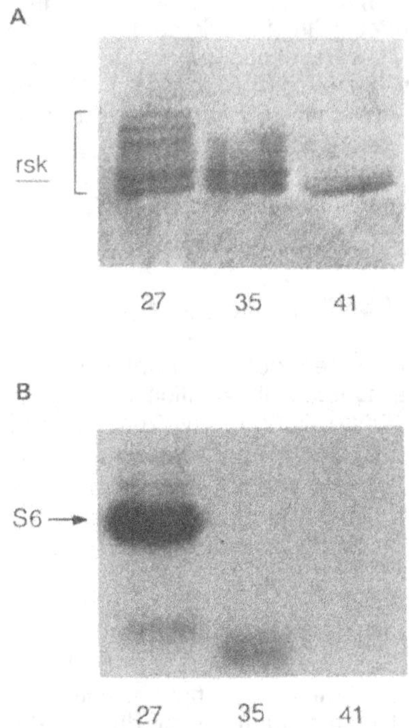

FIG. 3. Separation of active from inactive pp90. S-Sepharose column fractions were assayed by rsk immunoblot (A) and 40S phosphorylation activity (B). Insect cells were infected with rsk-a and v-src baculoviruses. The proteins from a cell lysate were eluted from a 5ml S-sepharose column as described. One milliliter fractions were collected and representative fractions were analyzed. A.) Twenty microliters of representative fractions were resolved on 8% SDS/PAGE, transferred to nitrocellulose and immunoblotted with rsk anti-serum. B.) Ten microliters were used for assaying direct 40S phosphorylation. The S6 band and rsk forms are indicated.

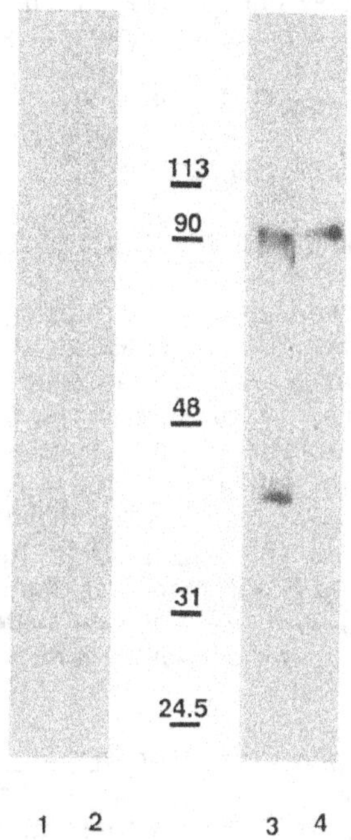

FIG. 4. Immunoblot of Xenopus S6KII and CEF Cell-free lysate. Samples of purified Xenopus S6KII (lanes 2 and 4) and serum-stimulated CEF cell-free lysate (lanes 1 and 3) were fractioned on 10% SDS-polyacrylamide gels (17) and immunoblotted either with preimmune serum (lanes 3 and 4). Positions of the molecular weight standards ($x10^{-3}$) are indicated.

As shown in Fig. 5, the in vivo phosphorylation of pp90 was increased by treatment with serum (lane 4), PMA (lane 6), or EGF/insulin (lane 8) as compared to untreated cells (lane 2). Mitogenic stimulation had no effect on the pattern or extent of phosphorylation observed for samples immunoprecipitated with non-immune serum (odd-numbered lanes). In parallel experiments, immunocomplex phosphorylation of pp90 and S6 was elevated similarly by these mitogens.

The same biosynthetically-labeled CEF lysates were used to identify the specific amino acids phosphorylated in pp90 in response to mitogen treatment. As shown in Figure 6, phosphoserine was the predominate phosphoamino acid in pp90 immunoprecipitated from quiescent cells treated for 1 hour with serum (panel B), PMA (panel C) and EGF/insulin (panel D). Stimulation with each of these mitogens also resulted in the phosphorylation of pp90 on threonine. It should also be noted that serine and threonine were the only phosphorylated amino acids detected in cells after only 10 minutes of stimulation with serum (data not shown). In extracts from CEF with NY72-4 and grown at 41°C, serine and threonine were the only phosphorylated amino acids detectable in pp90 (panel E). When NY72-4-infected cells were shifted to 35°C, phosphotyrosine, in addition to phosphoserine and phosphothreonine, could be detected in the immunoprecipitated pp90 (panel F).

The rsk gene(s) is likely to encode more than one protein kinase

The initial rsk cDNA clone was obtained using oligonucleotide probes with squences based on peptides obtained from purified S6KII. In the screening of tha cDNA library a number of related clones were identified, although only two were sequenced. These both predicted very similar protein sequences; however, as shown in Table I, only 4 of the 7 peptides with unambiguous sequences were predicted by these clones. We interpret this to mean that the purified enzyme is a mixture of different molecular species and that additional cDNAs remain to be identified. It should also be noted that S6KI is closely related to S6KII[18] and that the polyclonal antibody 125 immunoprecipitates both S6KI and S6KII. Moreover, there are at least two rsk genes in the murine genome, but to date we have made no systematic attempt to detect other related genes in any species. Thus, we cannot be certain whether we have studied S6KI or S6KII in mitogen-stimulated cultured cells.

Two enzymes contribute to S6 phosphorylation in serum-stimulated cells

Recently, Susa and colleagues have demonstrated that stimulation of S6 kinase activity in Swiss 3T3 cells activated with epidermal growth factor (EGF) follows a biphasic time course[19]. There is an early phase of activity at 10-15 min, followed by a short period of decline, then a late prolonged phase beginning after 30-60 min. The authors ascribe these results to the differential activation of a single enzyme, presumed to be the 70,000 M_r protein previously purified from Swiss 3T3 cells. We have also recently observed a time-dependent, biphasic activation of S6 kinase activity in CEF in response to serum-stimulation and have shown, using the polyclonal antiserum to Xenopus S6KII, that $pp90^{rsk}$ apparently contributes to the early phase (5-15 min) of the serum response and, to a lesser extent, to the later phase.

Figure 7 illustrates this time-dependent, biphasic increase in S6 kinase activity in serum-stimulated quiescent CEF as measured by phosphorylation of 40S subunits in vitro. The total S6 kinase activity as measured in cell-free lysates showed an initial decrease detected within 5 min of serum addition. This initial phase of activity was maximal at 15 min and decreased by 30 min. The second, prolonged phase of serum-stimulated activity was apparent at 1 h and remained significntly elevated even afer 8 h of serum treatment. Comparable levels of maximal S6 phosphorylation were detected during both phases of serum-stimulation. As described above, polyclonal antiserum 125, raised against recombinant Xenopus S6KII, immunoprecipitates a serum-

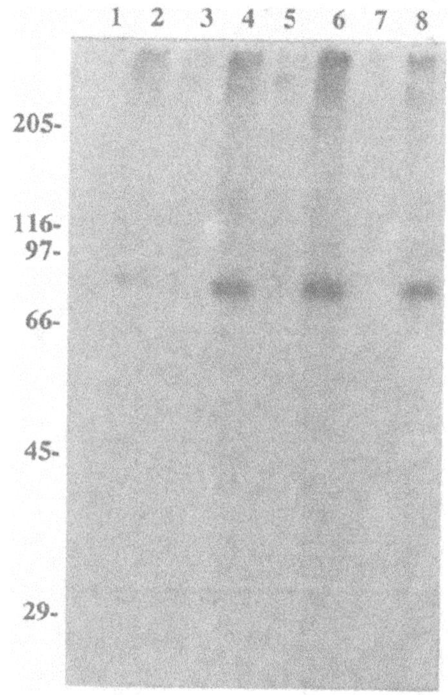

FIG.5. Immunoprecipitation of pp90 S6 kinase from biosynthetically-^{32}P labeled CEF. CEF were starved for phosphate and serum for 18 h prior to incubation for 1 h with addition of $H_3{}^{32}PO_4$ alone (lanes 1 and 2) or $H_3{}^{32}PO_4$ and dialyzed 10% calf serum (lanes 3 and 4), 100 ng/ml PMA (lanes 5 and 6), or EGF/insulin (lanes 7 and 8). One hour after ^{32}P addition, the cells were harvested, solubilized by boiling in 1% SDS, and immunoprecipitated with control serum (odd-numbered lanes) or 125 antiserum (even-numbered lanes). Immunoprecipitates were resolved on 10% SDS-polyacrylamide gels and subjected to autoradiography for 72 h. The migration of molecular weight standards ($x10^{-3}$) is indicated.

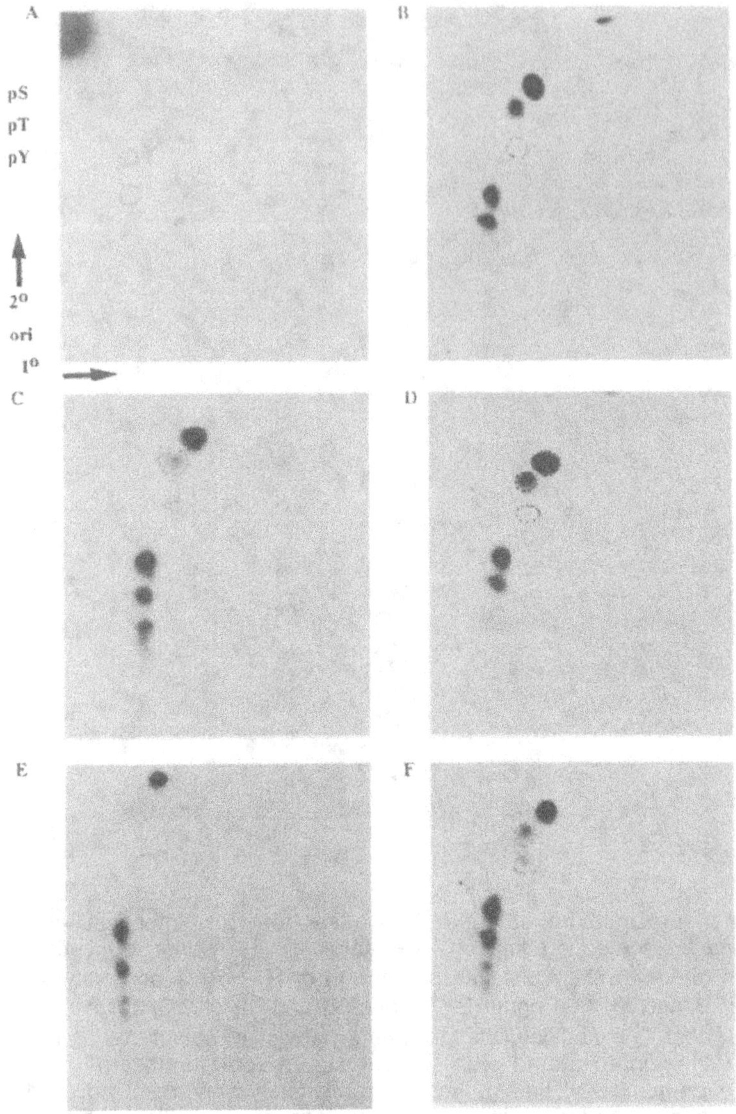

FIG.6. Phosphoamino acid analysis of pp90. CEF were biosynthetically 32P-labeled as described in Figure 5, and incubated for 1 h with either 10% dialyzed calf serum, 100 ng/ml PMA, or EGF/insulin. The positions of phosphoamino acid standards, as visualized by ninhidrin, and the origin are indicated. NY72-4-infected CEF were maintained at 41°C or shifted to 35°C for 24 h prior to labeling. Autoradiograms of the phosphoamino acids resolved by two-dimensional thin layer electrophoresis are presented as follows: A, control; B, 10% dialyzed calf serum; C, 100 ng/ml PMA; D, EGF/insulin; E, NY72-4, 41°C; F, NY72-4, 35°C.

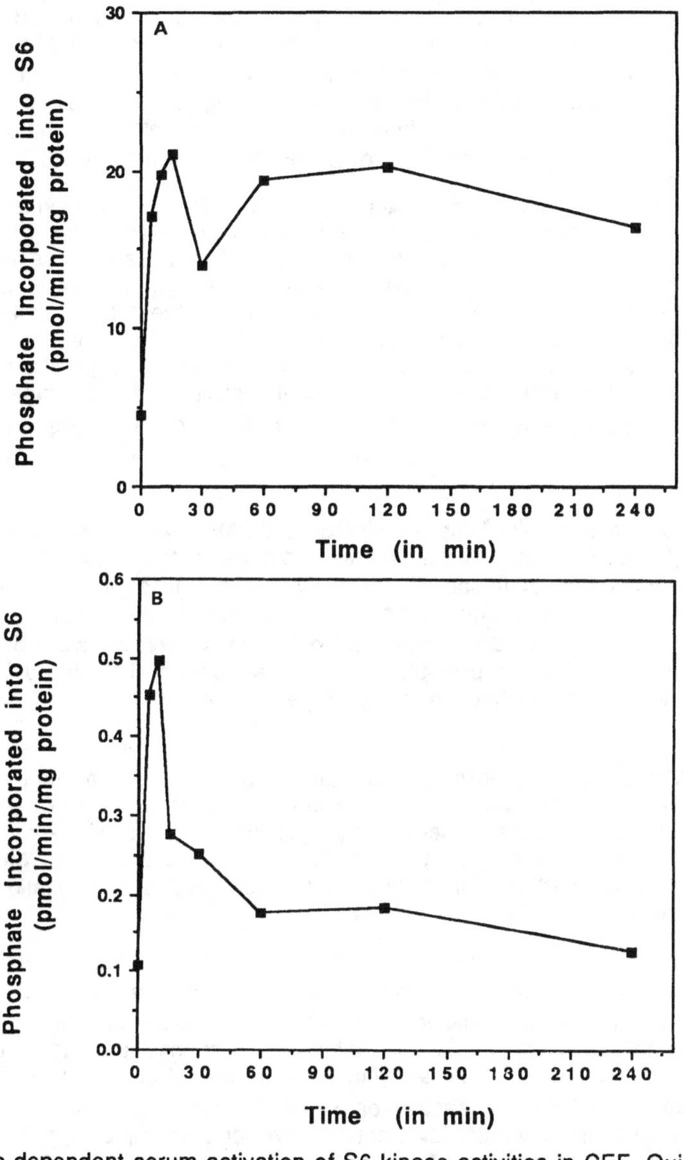

FIG. 7. Time-dependent serum activation of S6 kinase activities in CEF. Quiescent CEF were incubated with serum for the indicated times before the preparation of cell-free lysates. Lysates were adjusted to equal protein concentration and were assayed (A) directly (25mg lysate protein/5ml) or (B) following immunoprecipitation with antiserum 125 (250 mg lysate protein/50ml) for S6 kinase activity. The reaction mixtures were resolved by SDS-PAGE. The radiolabeled S6 band was identified by autoradiography, excised, and counted in liquid scintillant.

stimulated 90,000-M_r S6 kinase, pp90^{rsk}, from CEF[12,13]. This antiserum was used to compare the kinetics of serum-stimulation of pp90^{rsk} and total cellular S6 kinase activity. As was observed for total cellular S6 kinase activity, serum stimulated a rapid increase in S6 phosphorylation by pp90^{rsk}. This activity, measured in immunocomplexes, was increased at 5 min, maximal at 10 min, and significantly decreased at 30 min (Figure 7B). During the later, prolonged phase of kinase activation, immunoprecipitable pp90^{rsk} activity was reduced to about 30% of the maximum immunoprecipitable activity observed during the initial phase, but did not return to basal levels. In separate immunocomplexes assays, autophosphorylation of pp90^{rsk} paralleled the phosphorylation of S6 protein (data not shown).

To further analyze the kinase activities contributing to the biphasic stimulation of S6 phosphorylation, a cell lysate from CEF stimulated for 15 min (Figure 8A) was fractioned by ion-chromatography. S6 kinase activity was resolved as two peaks on DEAE-Sephacel, eluting at 85 mM NaCl (peak 1) and 175 mM NaCl (peak 2), with greater than 80% recovery of the total activity applied to the column. In Figure 8B, we show a comparative analysis and fractionation of lysates from cells that have been stimulated for 60 minutes. The results show peak 1 contributes a greater percentage of total activity at the early time points. Immunoprecipitable S6 kinase activity of pp90^{rsk} could be detected only in peak 1, independent of the duration of serum-stimulation.

The S6 kinase activity initially isolated as DEAE peak 1 was further purified to yield a single peak coincident with an autophosphorylated 90,000-M_r protein (data not shown). Immunoprecipitation of this activity with antiserum 125 identified it as pp90^{rsk}. An autophosphorylated 65,000-M_r protein co-eluted with the S6 kinase activity initially detected in DEAE peak 2..No S6 kinase activity could be detected in these fractions by immunoprecipitation with antiserum 125 (data not shown), consistent with the results obtained for DEAE peak 2 fractions.

To address the question of the structural relationship of the immunoprecipitable 90,000-M_r protein and the non-immunoreactive 65,000-M_r protein, phosphopeptide maps were generated by partial proteolysis with elastase and V8 protease. The phosphopeptide patterns indicated that the 90,000 M_r and 65,000 M_r proteins isolated acccording to the protocols described in this comunication are distinct (data not shown).

Thus, we conclude that at least two distinct enzymes are activated after serum-stimulation of cells that are able to contribute to S6 protein phosphorylation[12]. The activation of total cellular S6 kinase activity in response to serum-stimulation is biphasic and appears to be consistent with the activation of multiple S6 protein kinases with different activation and inactivation kinetics. A rsk gene product, pp90^{rsk}, is rapidly activated to its maximum level between two and ten minutes, followed by a fall in activity to a level betweeen 15 and 40% of the maximum. By 30 minutes after stimulation, a 65,000 M_r kinase that is unrelated to pp90^{rsk} apparently contributes a major fraction of the total activity measured in direct assays. Previous attempts at purification of S6 phosphorylating activity from CEF serum-stimulated for 30 min to 2 h resulted in the identification and characterization of a single 65,000-M_r S6 kinase, a protein presumably homologous to the 70,000 M_r protein purified from mammalian sources. The polyclonal antiserum 125, raised against recombinant Xenopus S6KII, allowed the identification of a serum-stimulated 90,000-M_r CEF homolog of S6KII, pp90^{rsk}. In addition, this antiserum immunoprecipitates a 90,000-M_r S6 kinase from NIH 3T3 and Swiss 3T3 cells (our unpublished results and J. Blenis, personal communication). These results suggest that the S6 protein

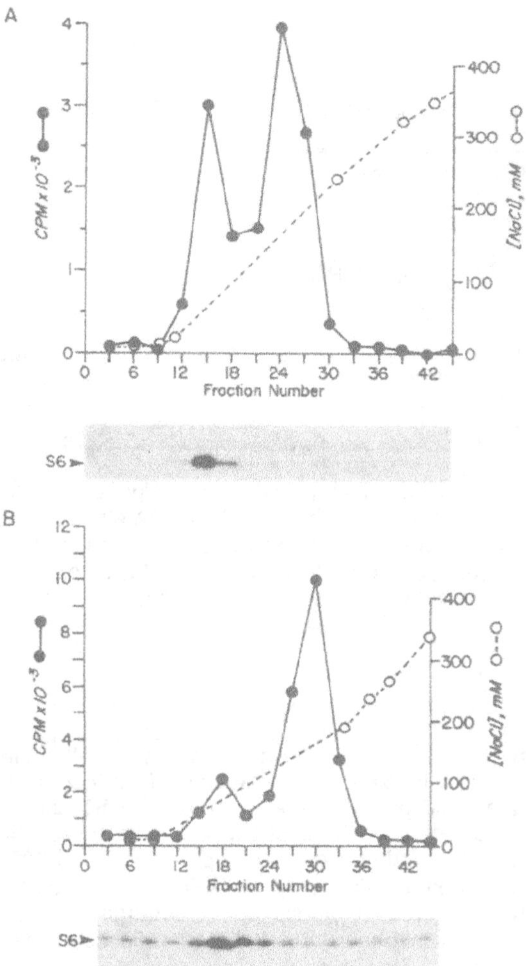

FIG. 8. Resolution of S6 kinase activities by DEAE-Sephacel chromatography. A.) Cell-free lysates prepared from approximately 1.5 x 10^{10} CEF 10% serum for 15 min, were resolved on a 4 ml DEAE-Sephacel column. Fractions were assayed for total (upper panel) S6 kinase activity. Reaction mixtures were resolved as in Figure 7. Immunoprecipitable S6 kinase activity (lower panel) in the same fractions was assayed and resolved by SDS-PAGE. The portion of the autoradiogram showing the radiolabeled S6 is presented. B.) CEFs were serum-stimulated for 60 minutes before chromatography and the fractions were assayed for immunoprecipitable S6 kinase activity.

kinase activity measured in total cell extracts prepared at various times is the sum of the activity of two enzymes that are likely to be activated by different pathways. A number of experiments suggest that both enzymes are regulated by reversible phosphorylation; therefore, the protein kinases and phosphoprotein phosphatases involved in the respective pathways are of considerable importance.

TABLE 1

Peptide Sequences from S6 Kinase II

Peptide	Sequence Obtained	
	Matched	
85-3	LTDFGLSK	
89-5	ICDFGFAK	
96-2	DLKPSNILYVDESGNPESIR	
96-3	DLKPENILLDEEGHIK	
	Not Matched	Closest Match Predicted
84-2	ISGTDAGQLYAMK	ItppDAnQLYAMK
89-6	ADPSQFELLK	ADqSdFvLLK
99-6	ADPSHFEFLK	ADqSdFvlLK

The peptides from S6 kinase II that yielded unambiguous sequence data are shown. Sequences that are predicted by one or more of the cDNA clones are given and are boxed in Fig.3. Peptides 84-2, 89-6, and 99-6 were not predicted. The closest match observed for the Xenopus clones is shown, with the lower case letters indicating the unmatched amino acids found in the predicted sequences.

SUMMARY

During the past several years we have attempted to determine how the activation of tyrosine-specific kinases leads to the activation of serine-specific protein kinases. We have selected as an end point the ribosomal protein S6 and studied the enzymes responsible for its phosphorylation. At least two enzymes participate in this event, one of approximate 90,000 M_r and the other of approximate 70,000 M_r. The former is the product of a gene we denote rsk. This gene is likely to encode several members of a protein kinase family. The product of rsk, pp90, undergoes a remarkable degree of activation upon expression in baculovirus-infected cells coinfected with recombinant virus expressing pp60$^{v\text{-}src}$. We and others have shown, however, that neither S6 kinase appears to be phosphorylated on tyrosine residues after mytogenic stimulation. Since pp90^{rsk} clearly seems to be regulated by reversible phosphorylation, another protein kinase transduces the signal to pp90^{rsk} from pp60$^{v\text{-}src}$. Thus, there are at least two S6 kinases that are likely to be activated by independent pathways, and we are in the process of purifying interesting potential candidates for these functions. There is also a strong possibility that these enzymes may have other physiologically significant substrates, and we have initiated studies in this research area as well.

REFERENCES

1. L.B. Ray and T.W. Sturgill, Insulin-stimulated microtubule-associated protein kinase is phosphorylated on tyrosine and threonine in vivo, Proc. Natl. Acad. Sci. USA 85: 3753 (1988).

2. E.G. Krebs, R.N. Eisenman, E.A. Kuenzel, D.W. Litchfield, F.J. Lozeman, B. Luscher, and J. Sommercorn, Casein kinase II as a potentially important enzyme concerned with signal transduction, Cold Spring Harbor Symposia on Quantitative Biology 53: 77 (1988).
3. P. Nurse, Universal control mechanism regulating onset of M-phase, Nature (London) 344: 503 (1990).
4. D.K. Morrison, D.R. Kaplan, U. Rapp and T.M. Roberts, Signal transuction from membrane to cytoplasm: Growth factors and membrane-bound oncogene products increase Raf-1 phosphorylation and associated protein kinase activity, Proc. Natl. Acad. Sci. USA 85: 8855 (1988).
5. L.M. Ballou, M. Siegmann, and G. Thomas, S6 kinase in quiescent Swiss mouse 3T3 cells is activated by phosphorylation in response to serum treatment, Proc. Natl. Acad. Sci. USA 85: 7154 (1988).
6. J. Blenis, C.J. Kuo and R.L. Erikson, Identification of a ribosomal protein S6 kinase regulated by transformation and growth promoting stimuli, J. Biol. Chem. 262: 14373 (1987).
7. D. Tabarini, A.G. de Garcia, J. Heinrich, and O.M. Rosen, Purification of a bovine liver S6 kinase, Biochem. Biophys. Res. Comm. 144: 891 (1987).
8. N.G. Anderson, J.L. Maller, N.K. Tonks and T. W. Sturgill, Requirement for integration of signals from two distinct phosphorylation pathways for activation of MAP kinase, Nature 343: 651 (1990).
9. D.K. Morrison, D.R. Kaplan, J.A. Escobedo, U.R. Rapp, T.M. Roberts and L.T. Williams, Direct activation of the serine/threonine kinase activity of Raf-1 through tyrosine phosphorylation by the PDGF b-receptor, Cell 58: 649.
10. E. Erikson and J.L. Maller, A protein kinase from Xenopus eggs specific for ribosomal protein S6, Proc. Natl. Acad. Sci. USA 82: 742 (1985).
11. T.A. Vik, L.J. Sweet and R.L. Erikson, Coinfection of insect cells with recombinant baculoviruses expressing $pp60^{v\text{-}src}$ results in the activation of a serine-specific protein kinase, $pp90^{rsk}$, Proc. Natl. Acad. Sci. USA 87: 2685 (1990).
12. L.J. Sweet, D.A. Alcorta, S.W. Jones, E. Erikson and R.L. Erikson, Identification of a mitogen-responsive ribosomal protein S6 kinase in chicken embryo fibroblasts, $pp90^{rsk}$: A homolog of Xenopus S6 kinase II, Mol. Cell. Biol. 10: 2413 (1990).
13. L.J. Sweet, D.A. Alcorta and R.L. Erikson, Two distinct enzymes contribute to biphasic S6 phosphorylation in serum-stimulated chicken embryo fibroblasts, Mol. Cell. Biol. (in press, 1990).
14. T.W. Sturgill, L.B. Ray, E. Erikson, and J.L. Maller, Insulin-stimulated MAP-2 kinase phosphorylates and activates ribosomal protein S6 kinase II, Nature (London) 334: 715 (1988).
15. S.W. Jones, E. Erikson, J. Blenis, J.L. Maller and R.L. Erikson, A Xenopus ribosomal protein S6 protein has two apparent kinase domains that are similar to distinct protein kinases, Proc. Natl. acad. Sci. USA 85: 3377 (1988).
16. D.A. Alcorta, C.M. Crews, L.J. Sweet, L. Bankston, S.W. Jones and R.L. Erikson, Sequence and expression of chicken and mouse rsk: homologs of Xenopus ribosomal S6 kinase, Mol. Cell. Biol. 9: 3850 (1989).
17. E. Erikson, D. Stefanovic, J. Blenis, R.L. Erikson, and J.L. Maller, Antibodies to Xenopus egg S6 kinase II recognize S6 kinase from progesterone- and insulin-stimulated Xenopus oocytes and from proliferation chicken embryo fibroblasts, Mol. Cell. Biol. 7: 3147 (1987).
18. E. Erikson and J.L. Maller, In vivo phosphorylation and activation of ribosomal S6 kinases during Xenopus oocyte maturation, J. Biol. Chem. 264: 13711 (1989).
19. M. Susa, A.R. Olivier, D. Fabbro, and G. Thomas, EGF induces biphasic S6 kinase activation: Late phase is protein kinase C-dependent and contributes to mitogenicity, Cell. 57: 817 (1989).

MOLECULAR BIOLOGY OF LEUKEMIAS AND LYMPHOMAS

Carlo M. Croce and Charlotte E. Gauwerky

Fels Institute for Cancer Research and Molecular Biology
Temple University School of Medicine
3420 North Broad Street
Philadelphia, Pennsylvania 19140

INTRODUCTION

Since the discovery of the Philadelphia chromosome as a consistent marker for the malignant cells in chronic myelogenous leukemias (1), non random chromosomal changes have been found in several neoplastic diseases of the hematopoietic system (2,3). Many of these somatic genetic changes involve reciprocal chromosomal traslocations (3), and some have now been shown to occur at or near a proto-oncogene locus. The presence of these chromosomal aberrations have suggested fruitful approaches to the molecular understanding of these tumors. We, therefore, have taken advantage of lymphoid-specific chromosome abnormalities to analyze the mechanisms of oncogenesis in B and T cells on the molecular level.

In 1972 Manolov and Manolova (4) described a 14q+ chromosome in Burkitt's lymphoma, a neoplastic condition of B-cells affecting predominantly children. Three reciprocal traslocations have been detected (4,5). Approximately 80% of them exhibit a t(8;14)(q24;q32) translocation, while the remaining 25% of the cases carry a t(2;8)(p11;q24) or a t(8;22)(q24;q11) translocation (5). The precise chromosomal bands at which these translocations take place is of considerable significance, because genes of central importance to B-cell ontogeny map to these chromosomal regions. Interestingly, the loci for human immunoglobulin heavy, lambda and kappa chain genes are located on human chromosome 14 (6), 22 (7), and 2 (8,9), respectively, which strongly suggests a relationship between human immunoglobulin genes, chromosome rearrangement and B-cell ontogeny. The immunoglobulin gene regions have been proposed to provide a cellular proto-oncogene with promoters active in B-cells (10). DNA sequences that function by promoting gene transcription may be regulated in a tissue specific manner. The immunoglobulin heavy chain locus maps to chromosome 14q32. The immunoglobulin k and l light chain genes map to chromosomes 2p11 and 22q11, respectively. Each of these chromosomal regions is a site of chromosome translocation in Burkitt's lymphomas. The other involved chromosome in each of these chromosomal aberrations is chromosom 8 at band q24. This is the site at which the human homolog of the avian myelocytomatosis virus transforming gene, c-myc, is located (11).Thus, cytogenetic evidence alone suggests that the translocations in Burkitt's lymphoma juxtapose the c-myc gene and immunoglobulin loci.

Molecular Basis of Human Cancer
Edited by C. Nicolini, Plenum Press, New York, 1991

Another B-cell malignancy commonly exhibiting a non-random chromosome translocation is follicular lymphoma (12). Follicular lymphoma is a low grade B-cell malignancy and the most common hematopoietic neoplasm in North America and Western Europe. In over 85% of the cases the malignant cells carry a t(14;18)(q32;q21) translocation - and the site of the immunoglobulin heavy chains locus is implicated in this translocation. This translocation involves a gene called bcl-2, initially identified in our laboratory through the molecular analysis of these translocations (12).

Finally, the t(11;14)(q13;q32) translocation is found in chronic lymphocitic leukemia, multiple myeloma and in diffuse smalll and large-cell lymphomas. In this translocation the immunoglobulin heavy locus on chromosome 14 and a locus termed bcl-1 on chromosome 11 are involved in this translocation (2,13).

T-cell malignancies are much less common than B-cell neoplasms, and they often exhibit chromosome aberrations involving band 14q11, which is the cytogenic location of the TCR alpha chain (14). Embedded within the TCR a is the gene for TCR d (15). Both of these genes are involved in several translocations. In the t(8;14)(q24;q11) translocation, the TCR a locus is joined 3' of c-myc (16). Other t(8;14)(q24;q11) translocations, involve the TCR d gene (17). In both T-cell and B-cell malignancies rearranging immunoglobulin superfamily genes often involved in translocations.

More recently, we have included into our studies investigations on tumor progression in B-cell neoplasm following Peter Nowell's concept that tumor progression is a multiple step event involving several subsequently occurring cytogenetic changes (18). These studies indicate that during the course of the disease in low grade B-cell malignancies which carry a t(14;18) translocation involving bcl-2 c-myc becomes activated in an additional translocation involving chromosome 8 (19). These studies recently resulted in the discovery of the bcl-3 gene which had joined into c-myc and activated it in a silent additional t(8;17) translocation (20).

In this review, I will describe how the methods and concepts of molecular genetics were used to characterize the precise breakpoints of the chromosomal translocations found in B-cell and T-cell malignancies as well as the genes activated in those neoplasms.

MOLECULAR GENETICS OF BURKITT'S LYMPHOMA

Burkitt's lymphoma is one of the better understood malignancies in molecular terms. Initially, our laboratory has used somatic cell hybridization techniques to determine whether the chromosomal breakpoints in Burkitt's lymphomas with the t(8;14) translocation would directly involve the immunoglobulin heavy chain locus (21) Hybrid cells between mouse and human Burkitt lymphoma cells that retained the 14q+ chromosome were found to contain the human heavy chain constant region (CH) genes, while hybrid cells retaining the 8q- chromosome contain the immunoglobulin heavy chain locus in the t(8;14) translocation in Burkitt's lymphoma. The same approach was used to elucidate the organization of the Burkitt's lymphoma variants with the t(2;8) and the t(8;22) translocation. Mouse x human hybrid cells formed with Burkitt's cells carrying the t(8;22) translocation revealed that the 22q- chromosome contained the Vl genes, while the 8q+ chromosome contained the Cl gene,indicating that the chromosomal break splits the lambda chain locus (22). Similarly, the break in the t(2;8) translocation separates the Vk and Ck translocates to the involved chromosome 8 (23) (Fig. 1).

We then found in collaboration with Riccardo Dalla-Favera that the human homologue (c-myc) of v-myc is located on the segment of chromosome 8 that is translocated to chromosome 14 in Burkitt's lymphomas with the t(8;14) translocation (11). Like the avian leukosis viruses, the avian myelocytomatosis virus oncogene (v-myc) can transform normal B-cells into lymphoma cells by integrating into the proximity to the normal c-myc homologue (24).

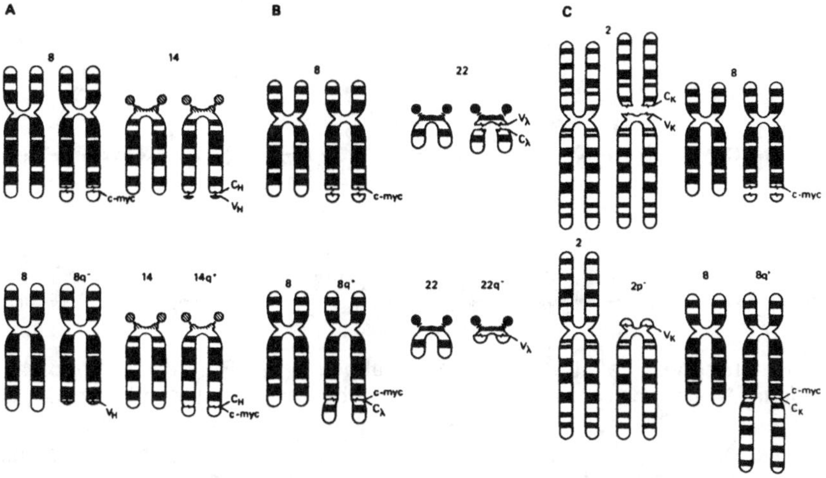

FIG.1 In Burkitt's lymphomas with the t(8;14) translocation, the c-myc oncogene tranlocates to the heavy chain locus (A), and a portion of the immunoglobulin locus (V_H) is translocated to chromosome 8. In Burkitt's lymphomas with the les frequent (8;22) (B) and t(2;8) (C) translocations, the c-myc oncogene remains on the involved chromosome 8, but the genes for the immunoglobulin light chain constant regions (Ck and Cl) translocate to a region 3' (distal) the c-myc on the involved on chromosome 8 (8q+). Again, with these translocations, the immunoglobulin loci are split so that sequences that encode for the variable portion of the immunoglobulin molecule Vk or Vl remain on chromosome 2 or 22, respectively.

We and others analyzed many cases of Burkitt's lymphomas. Figure 2 summarizes the structures of the various types of translocations, the most common (80%) involve c-myc and the Ig heavy chain locus. A general feature of the translocated chromosome is that the breakpoints of the t(8;14) translocation are heterogeneous. On chromosome 14 (Fig. 2) they are all situated at sites where normal, physiologic rearrangement occurs. As we know immunologic diversity is generated by shuffling variable , diversity and joining segments carried out by the V-D-Jjoining system of recombination enzymes (25). The other process of isotype switching allows for the diversification of the immune response in terms of effector function - which is carried out by switching enzymes. Both the recombination and switching enzymes act at specific sites in the IgH locus, precisely those where translocation occur. Thus we found that on chromosome 14 the breaks occur in the regions of V_H, D_H, J_H genes as well as in the switch regions of Cm, Cg and Ca. (Sm; Sg and Sa).

The break of chromosome 8 are also dispersed. In Burkitt's lymphomas with the t(8;14) chromosome translocation they are always at the 5' end of the coding sequence of the c-myc gene and occur either 5' to the gene or within the gene between the non-coding first exon and the second exon.

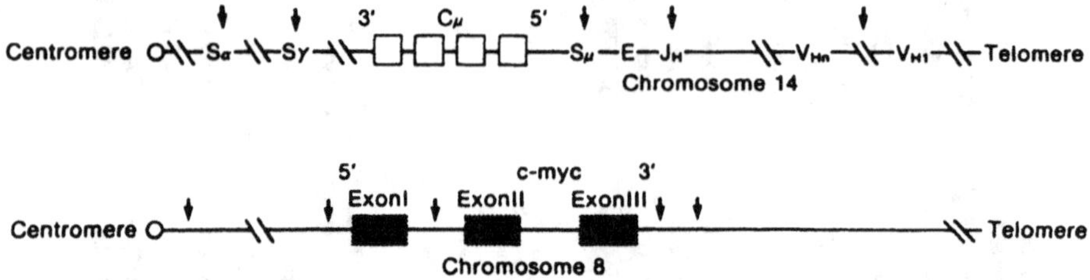

FIG.2 Heterogeneity of translocation breakpoints in Burkitt's lymphomas. On chromosome 14 (top) chromosome breakage has been demonstrated in Ca, Cg, Cm,J_H and D_H regions, as indicated by arrows. On chromosome 8 (bottom), breakpoints are 5' of c-myc or within the first intron in cases carrying t(8;14) translocations. The variant translocations resulted in breakage 3' of an intact c-myc.

These variations of breakpoints on chr 14 and on chromosome 8 has made us to distinguish endemic or so called African Burkitt's lymphomas from sporadic Burkitt's lymphomas (26) Table 1 summarizes the features of endemic and sporadic Burkitt's lymphomas. Endemic Burkitt's lymphoma occurs predominantly in Central Africa and those cases are Epstein Barr virus positive and the translocation on chr 8 occurs far 5' to the c-myc and on chr 14 in the variable, diversity or joining segment of the immunoglobulin heavy chain region. On the other hand, in sporadic Burkitt's lymphoma the break of the t(8;14) translocation involves directly c-myc and then, the broken c-myc gene joins into a switch region of the immunoglobulin heavy chain locus. The fact that endemic Burkitt does not secrete immunoglobulins let us suggest that different mechanisms are involved in the translocations in endemic and in sporadic Burkitt's lymphoma.

Our laboratory has studied the human genomic c-myc DNA and the entire c-myc cDNA, and we could identify that the c-myc oncogene is formed by three separated exons (27,28) (Fig.3). The first exon contains termination signals in all three reading frames and represents an untranslated leader sequence. At the beginning of the second exon there is the first ATG (methionine) signal for protein synthesis. Then, there is a single open reading frame involving the second and third exon which encodes for a protein of 439 amino acids and 48, 812 kd molecular weight. C-myc myc has 2 promoters, and the transcripts initiate from 2 preferential initiation of transcription sites. When c-myc becomes rearranged in a chromosomal break either in the first intron or in the first exon, new criptic promoters become activated within the first intron. Since the DNA segment immediately 5' of the c-myc second exon contains termination codes on all three reading frames, the protein product of the rearranged gene is identical to the product of the germ line involved c-myc gene (27,28).

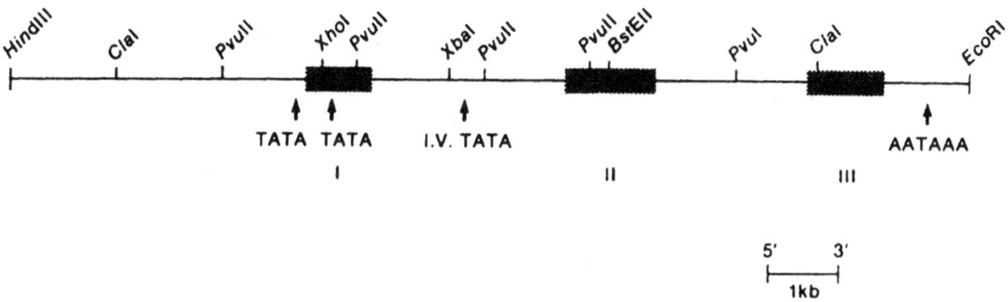

FIG.3 Schematic representation of the c-myc locus as determined by us by sequencing and S_1 nuclease analysis. Approximate location of the authentic TATA boxes found by us and others. "TATA" box like sequences (I.V. TATA) found within the first intron are indicated. Location of the recognition signal sequence (AAT AAA) for polyadenylation is also indicated.

We have studied the regulation of c-myc expression in Burkitt's lymphoma, and initially we found that the levels of c-myc transcripts in Burkitt's lymphoma are generally elevated but vary from case to case (29). In order to determine whether there is a difference in the expression of the translocated gene versus the normal c-myc oncogene, our laboratory has studied the expression of c-myc on a mouse myeloma background, and was assessing using S_1 nuclease analysis. We found that somatic cell hybrids containing the normal human chromosome 8 did not express human myc transcripts, while somatic cell hybrids containing the translocated c-myc gene on a chromosome 14q+ expressed high levels of myc transcripts indicating deregulation of the translocated c-myc gene in Burkitt's lymphoma with a t(8;14) translocation (29) (Fig. 4), and we further found that the translocated c-myc oncogene fails to respond to transcriptional control and is transcribed constitutively at high levels in these cells. Two basic hypothesis have been proposed to explain this deregulation of c-myc transcription after translocation. The first focuses on the role of trans-acting factor and their interactions with putative repressor binding sites in the 5' flanking region of c-myc (30); the other suggests the influence of cis acting elements in the Ig locus on c-myc expression(31,32).

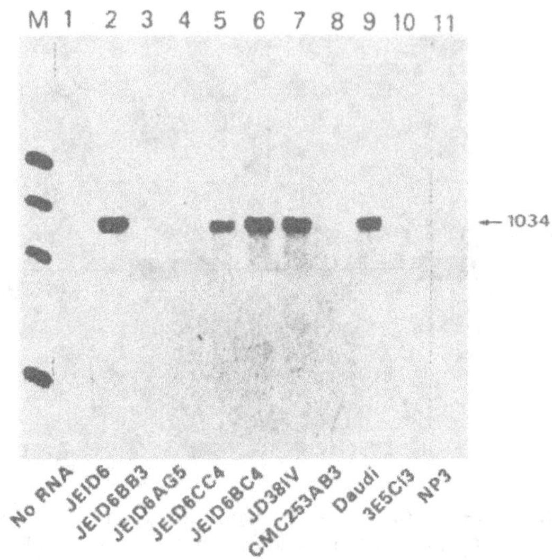

FIG.4 Nuclease S_1 analysis of c-myc RNAs with the hybrid cells between NP3 and Burkitt's lymphoma cell lines with the t(8;14) chromosomal translocation. Cytoplasmic RNA (20 mg) was hybridized with human c-myc probe. Parental NP3 used for hybrid preparation is a non-producer mouse myeloma. JE1D6 is a hybrid containing both the 14q+ and the normal chromosome 8.

Although it sems possible that translocations which separate the first exon from the rest of the c-myc gene may alter the regulation of c-myc transcription, several observations suggest that decapitation per se is not sufficient to explain deregulation. In gene transfection experiments only c-myc constructs containing either a viral or an immunoglobulin enhancer could be expressed (33,34). It therefore seems likely that it is the association with genetic elements capable of activating gene transcription in cis makes expression of myc constitutive, so the cells continue to proliferate without any restriction, leading to an aggressive B-cell malignancy.

AIDS-ASSOCIATED BURKITT'S LYMPHOMA

Burkitt's lymphoma also occurs in patients with the Acquired Immuno Deficiency Syndrome (35). These tumors share features with endemic Burkitt's lymphomas. Many of these occur in conjunction with infection by the Epstein Barr virus (36), while EBV is found in over 95% of endemic Burkitt's lymphoma cases; they arise in situations of severe systemic immunosuppression thought to be an important co-factor for the development of endemic Burkitt's lymphoma; and they are often preceded by expression of oligs clonal B-cell populations - an important step in the pathogenesis of endemic Burkitt's tumors (36). These clinical parallels between AIDS-associated and endemic Burkitt's lymphomas suggested to us that molecular similities might exist as well. We then have analyzed cases of AIDS-associated Burkitt's lymphoma where the malignant cells revealed a t(8;14) translocation and the patient was shown to carry HIV. In our studies the molecular analysis revealed rearrengement of the JH immunoglobulin heavy chain locus and germ line configuration for c-myc indicating that the break on chromosome 8 had occured far 5' of c-myc as in endemic Burkitt's and then it had directly involved in the heavy chain locus on chromosome 14 in the t(8;14) translocation. We then have used the probe specific for the JH region to clone the break in t(8;14) translocation in order to identify mechanisms involved in this translocation. Sequence analysis of the breakpoint region indicated the break had occurred nearly at the identical site previously described for the P3HR.1 endemic Burkitt's lymphoma breakpoint in the very 5' of c-myc by mechanisms involving the V-D-J enzyme machinery as in endemic Burkitt's lymphoma (37) (Fig.5).

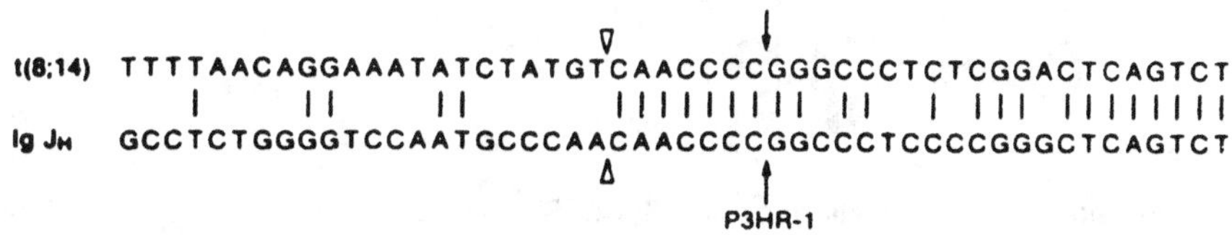

FIG.5 Nucleotide sequences of the AIDS-associated lymphoma t(8;14) breakpoint and the normal 14 J_H region. The breakpoint is indicated by D. This P3HRI breakpoint is indicated by ≠.

MOLECULAR GENETICS OF FOLLICULAR LYMPHOMA

A striking cytogenetic similarity between the Burkitt's lymphomas discussed above and several other lymphoid malignancies is the consistent involvement of chromosome 14q32 in chromosome translocations. Nearly 90% of follicular lymphoma exhibit a t(14;18)(q32;q21) translocation (2) (Fig. 6). These translocations uniformly involve the JH segments of the IgH locus. On chromosome 18, the translocation breakpoints are also consistently clustered and involved the locus of a transcriptional unit initially designated bcl-2 (38). Bcl-2 was cloned and characterized in our laboratory. This gene is located on chr 18 at band q21, and is involved in the t(14;18) translocation in follicular lymphoma. The region of chr 18 involved in the t(14;18) translocation was first cloned from the 380 cell line (39). The cell line was derived from a patient with acute lymphoblastic leukemia and carries the t(8;14) and t(14;18) translocations (39). By screening a genomic library made from 380 genomic DNA with JH probes, two sets of clones were obtained. These correspond to translocation breakpoints from both 14q+ chromosomes. Thus, both JH alleles in this neoplastic clone were involved in translocations. One of them had joined into the bcl-2 locus on chr 18q21.

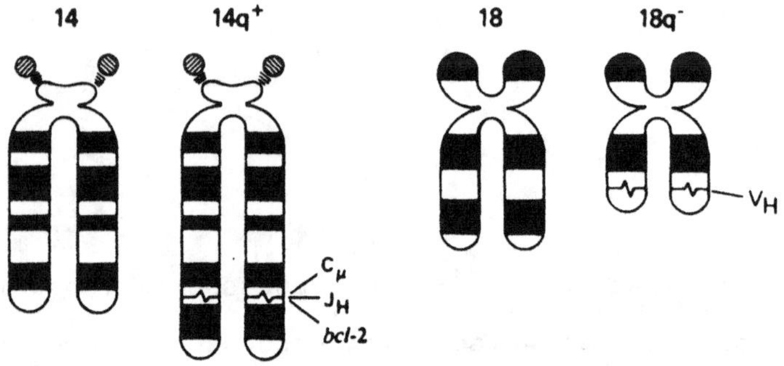

FIG.6 Schematic presentation of the t(14;18)(q32;q21) chromosomal translocation observed in follicular lymphomas. Note the juxtaposition of J_H and bcl-2 sequences on the 14q+ chromosome.

We then have used chromosome walking techniques and pulse field gel electrophoresis to characterize bcl-2. Bcl-2 is a very large gene. Its final lenght is not known yet, but data indicate that it might have a lenght of 400 Kb. Fig 7 gives us a rough idea on the architecture of bcl-2. The gene comprises 2 exons which are separated by a large intron of around 370 Kb. These are two distinct promoter regions. Exon I has classic TATA box plus CAAT box and consistent with this, two distinct initiation sites can be defined. Also a decanucleotide motif could be identified which has sequence homology to the Ig variable regions enhancers and which most likely serves as a tissue specific enhancer for this B-cell specific gene (40).

In addition, an entirely separate promoter region exists further 5' in exon I. There is a GC region with Sp1 binding motifs but there is no TATA box and correspondingly demonstrates multiple sites of transcriptions initiation.

The two bcl-2 exons are transcribed into three transcripts of 3.5, 5.5 and 8.5 Kb in lenght, respectively (41). These transcripts are, in turn, translated into two proteins, bcl-2a, and bcl-2b, which are 239 and 205 amino acids in lenght, respectively. The proteins differ only in their carboxil termini.

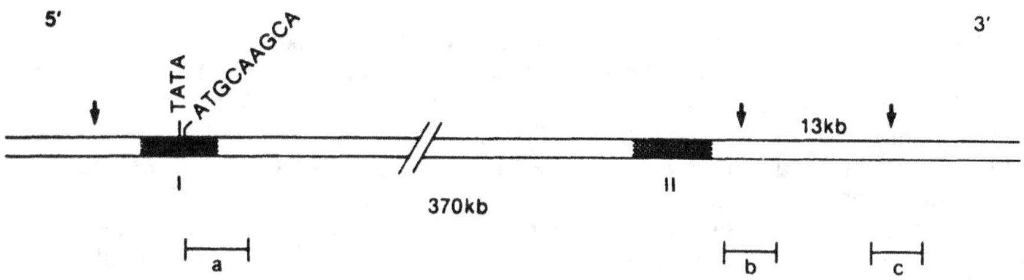

FIG.7 Schematic representation of the bcl-2 locus as described by us and others (40,41). Exon II is separated by a large intron of 370 kb from the first exon complex. The first exon contains two promoter regions, Sp1 binding sites and a decanucleotide motis with similarly to those found in immunoglobulin variable regions. Probes used to identify the major breakpoint cluster region in the 3' are indicated. Note the location of the breakpoint cluster region for CLL in the 5' of bcl-2 identified using probe C.

The expression of these proteins is influenced by mitogenic stimuli, and they are significantly conserved across species boundaries, strongly suggesting an important role for this gene in cell proliferation (42). The bcl-2 gene product is located at the inner surface of the cell membrane (43), it has GTP-binding activity, and a protein sequence that suggests it belongs to the small molecular weight GTP-binding protein (Gprotein) family (44). Thus, the bcl-2 product seems to function in mitogenic signal trasduction (44).

In parallel to our investigation in Burkitt's lymphoma in folicular lymphoma, translocation results in the deregulation of bcl-2 expression. Deregulation takes place as a consequence of enhanced bcl-2 transcription. It is likely that this occurs because of the abnormal proximity of the bcl-2 gene to enhancer elements within the IgH locus. Because the chromosome breakpoints all fall within the JH segments, the IgH enhancer, which lies immediately 3' of JH, is closely positioned to the chromosome junction. Thus, the consistent transcriptional deregulation of bcl-2 in the setting of follicular lymphoma strongly argues for the oncogenic potential of bcl-2 in B-cells (41).

We have developed different probes from the bcl-2 locus which are indicated in Fig.7 as probe A, probe B and probe C, and then we have analyzed the involvement of the different regions of the bcl-2 locus in follicular lymphoma. Using probe B from the 3' untranslated region of the 2 exon we were able to detect rearrangement in around 60% of all follicular lymphomas indicating the presence of a major breakpoint cluster region in this low grade B-cell neoplasm (12). When we used probe C from the more 3' of the bcl-2 locus we found rearrangement in approximately 20% of follicular lymphoma cases, giving evidence for a minor breakpoint cluster region in the 3' of bcl-2 (12). But we also found a single follicular lymphoma case with rearrangement in the 5' of bcl-2 which could be demonstrated using probe A (45). These findings suggest that breaks can occur throughout the entire bcl-2 locus and that our probes are appropriate to detect rearrangement in 80% of the cases.

More recently we have studied the involvement of bcl-2 also in another low grade B-cell malignancy, the chronic lymphocytic leukemia. First, in collaboration with Dr. Tsujimoto we found a case of CLL with a break in the 5' of bcl-2 where bcl-2 had joined into the immunoglobulin light chain region of Cl (46). In analogy to the Burkitt's lymphoma variants with the t(2;8) or t(22;8) translocation we called this t(18;22) translocation a bcl-2 variant. Later on Tsujimoto reported that the 5' of bcl-2 appears to be consistently involved in CLL in 10% of the cases not only involving Cl but also Ck on chromosome 2 (47). In collaboration with Daniel Catovsky in London we could detect rearrangement for the 5' of bcl-2 in CLL in ≈ 30% of the cases and we hope to be able to confirm this first preliminary observation by analyzing more cases (48).

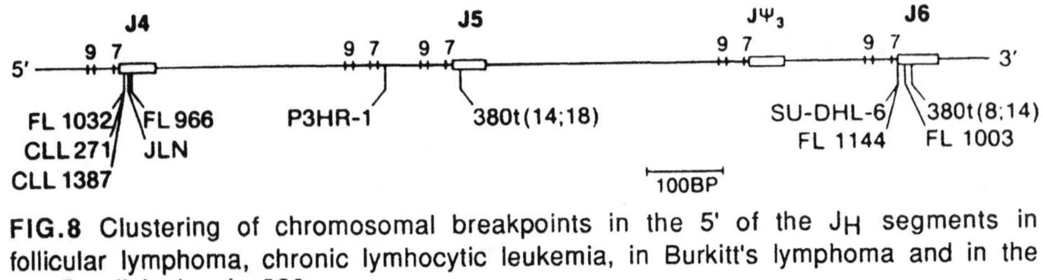

FIG.8 Clustering of chromosomal breakpoints in the 5' of the J_H segments in follicular lymphoma, chronic lymhocytic leukemia, in Burkitt's lymphoma and in the pre B-cell leukemia 380.

We have cloned the breakpoints for many different cases of follicular lymphoma, and we found that in every case the breakpoint involved the 5' end of a J region of the heavy chain. Different J_H could be involved and the heavy chain associated with the J could be a constant m, a constant g or a constant a. The clustering of breakpoints 5' of JH segments at the immunoglobulin heavy chain locus on chromosome 14 in follicular lymphoma is shown in Fig.8 (49).

MOLECULAR GENETICS OF THE t(11;14) TRANSLOCATION: THE bcl-1 BREAKPOINT

As mentioned above, a variety of tumors exhibit the t(11;14) translocation. This abnormality is found in multiple myeloma, and in chronic lymphocytic leukemia in 10-30% of the cases. Also this translocation involves the JH locus in the 5' region of J segments (Fig.8).

Breakpoints also cluster on chromosome 11 in CLL (50). We have termed the locus of this cluster region at chr 11q13 bcl-1. Also this locus has been explained extensively by chromosome walking, the transcriptional unit of the proposed gene bcl-1 has not been identified yet.

MOLECULAR BIOLOGY OF T-CELL MALIGNANCIES

The molecular organization of the T-cell receptor gene recently has been demonstrated to bear remarkable similarity to that of Ig genes. Variable, diversity and joining segments rearrange to form a full variable exon during early T-cell development, and transcription of the variable and constant exons follows. The TCR loci, like the Ig loci, are thus subject to abnormal recombination. Furthermore, the tissue specific TCR expression promotes activation of translocated oncogenes.

The t(8;24)(q24;q11) translocation is the best understood of the T-cell specific translocations. In this translocation the c-myc gene joins with the tcr a/d locus on chromosome 14 (51). Somatic cell hybridization studies initially demonstrated interruption of the TCRa locus by this translocation. Cloning and sequencing of the breakpoint region revealed that the 3' region of c-myc was translocated to the 5' end of one of the Ja segments. Furthermore, hybrids carrying the translocated myc gene expressed it inappropriately in a manner previously shown for Burkitt's lymphoma. These studies strenghten the analogy between chromosome translocations in T cells and in B cells by demonstrating that the TCR loci may function to translocation and activate oncogenes.

We have used the somatic cell hybridization technique to analyze other T-cell specific translocations. The TCRa locus is split between the Ca and Va segments in both the t(10;14) and the t(11:14) translocations (52,53). We recently found that the TCRd gene is embedded within the TCR Ja region, and we further could show that TCRd is involved in the t(10;14) (54) and in the t(11;14) (55) translocations, and we then hypothesized that genes important to T-cell oncogenesis - tcl-3 - and tcl-2 - are the involved portions of chromosome 10 and 11 respectively. However, transcriptional units have not been identified to date.

Another important T-cell specific chromosomal aberration involves chromosome 14q32, the location of IgH gene. Such T cell neoplasms carry either inv(14)(q11;q32) inversions or a t(14;14)(q11;q32) translocation (56). This locus at chr. 14q32 is also involved in the t(7;14) translocation found in ataxia telangiectasia (57). Our investigations suggest that a gene proximal to the IgH genes at chr. 14q32 must be involved; we previously designated this putative gene as tcl-1.

More recently we were able to characterize at the molecular level the t(1;14)(p32;q11) translocation, described to occur for acute T-cell leukemias. This translocation involves on chromosome 14 the region of the diversity segment Dd , of the TCR a/d locus, and on chromosome 1p32 a locus that is also involved in malignant melanoma and neuroblastoma (58). Cloning of this translocation resulted in the identification of a DNA segment at 1p32 that was able to identify a transcriptional unit which we called tcl-5 (59) (Fig.9).

MOLECULAR MECHANISMS OF CHROMOSOMAL TRANSLOCATIONS IN LYMPHOID MALIGNANCIES

There is now persuading evidence that the most likely mechanism leading to chromosomal translocations in T and B cell neoplasms are recombinase errors. Chromosomal breaks occur at sites of recombinase activity; i.e., in Burkitt's lymphoma the breaks are seen preferentially 5' of the DH or JH segments, or in switch regions (49). The t(11;14) and the t(14;18) breakpoints seen in CLL and in follicular lymphoma also cluster 5' of the JH segments on the 14q+ chromosome (50, 60); the 11q- and the 18q_ chromosomes exhibit DH involvement.

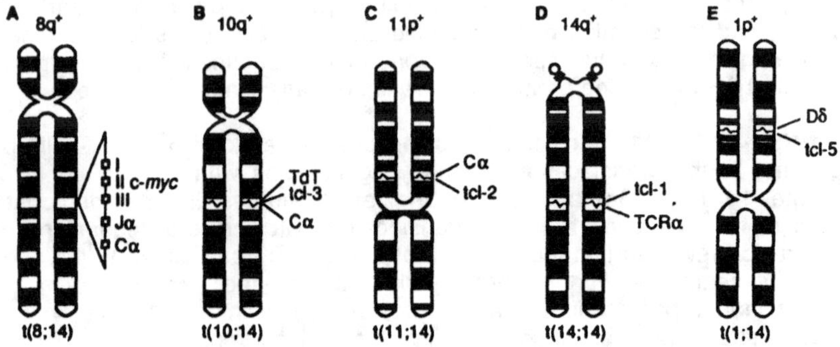

FIG.9 Translocations commonly observed in various T-cell leukemias and lymphomas. In A, the t(8;14)(q24;q11) translocation is illustrated; we have cloned the breakpoint from this tumor and demonstrated the juxtaposition of sequences 3' of myc and the TCR Ja segment. The location of putative T-cell oncogenes are illustrated.

In TCR translocations, the 5' ends of Ja segments or Db-Jb regions are involved (49). All of these sites are those located at which physiologic recombination occurs. Near the translocation breakpoints we find a nucleotide sequence with features that indicates VDJ recombinase activity. This had first been shown for braks involved in CLL with the t(11;14) chromosome translocation and in follicular lymphoma with the t(14;18) translocation. These features include the presence of N regions and heptamer-nonamer motifs near the break (16, 50, 60). N-regions are stretches of oligonucleotides which are added by the enzyme terminal deoxynucleotidyl transferase (TdT) (63). Heptamers and nonamers, which are separated by twelve nucleotides and which are found on both sites of the chromosomal break, are thought to be signals for the VDJ recombinase enzymes (25) (Fig.10).

These features are highly conserved among TCR genes and immunoglobulins. The presence of these signals at translocation breakpoints supports the concept that they play an important role in the process of translocations. We had demonstrated this for follicular lymphoma, and also found the same recombinase mechanism involved in the

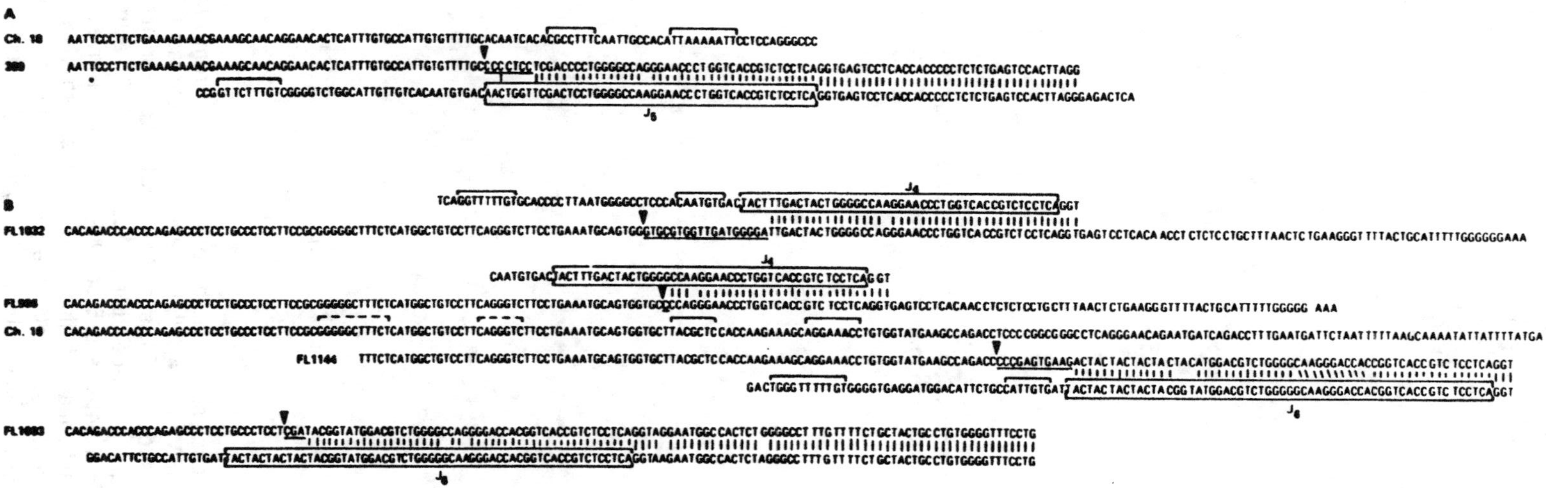

FIG.10 Sequences of breakpoints derived from samples carrying t(14;18) translocations. For each case, the position of the breakpoint is indicated by the arrow. Heavy chain J_H segments are bracketed. Heptamers-nonamers motifs are also bracketed, and N-regions are underlined.

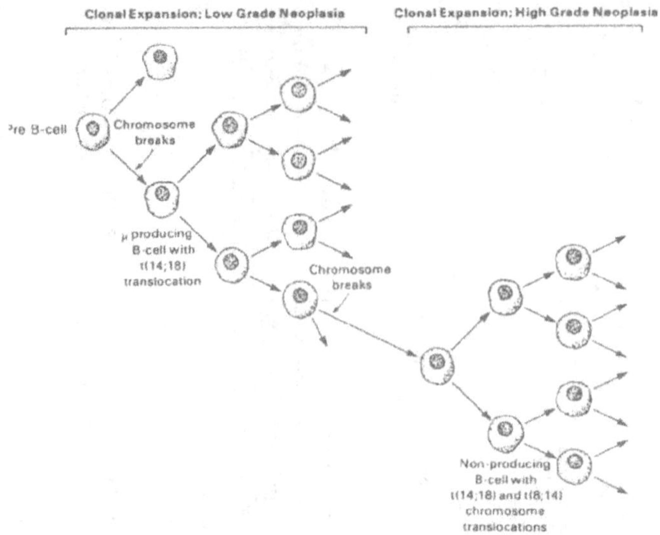

FIG.11 Model on tumor progression in B-cell neoplasms carrying two reciprocal translocations involving the two human chromosome 14, and chromosome 18 and 8, respectively.

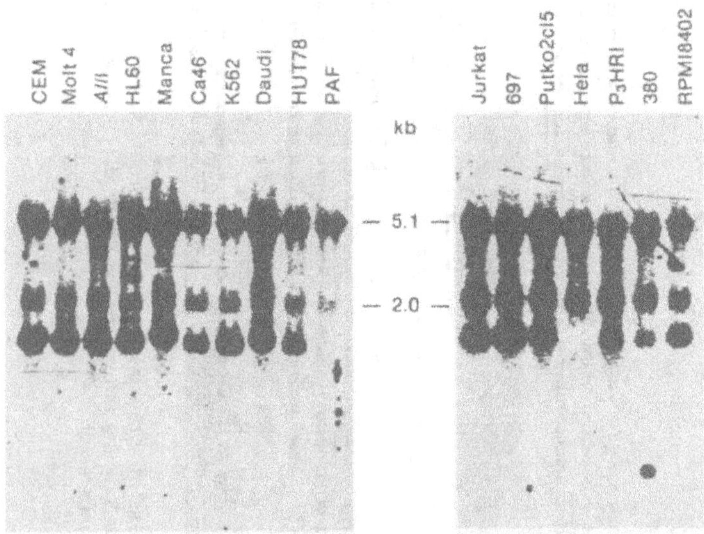

FIG. 12 Northern blot analysis of total cellular RNA from different human hematopoietic cell lines. CEM, Molt4, HUT78 and Jurkat represent T-cell leukemia cell lines. Manca, CA46, Daudi, P3HRI, and Putko 2Cl 5 are Burkitt's lymphoma cell lines, and 697 and 380 are pre B-cell lines. ALL I ia a PH+ acute lymphoblastic leukemia cell line, and RPMI 8402 is a B-cell line. HL60 is a myeloid leukemia cell line, K562 is an erythroleukemia cell line. PAF is a simian virus 40-transformed human kidney cell line, and HeLa cells are derived from a human carcinoma. Hybridization to a bcl-3 probe revealed the 1.7 transcript of the bcl-3 gene.

t(11;14) translocation in chronic lymphocytic leukemia and then in the t(8;14) translocation in endemic Burkitt's lymphoma. In addition, we find a similar mechanism involved in translocations in T-cell leukemias and lymphomas.

MOLECULAR EVENTS IN TUMOR PROGRESSION

More recently, we have studied the molecular genetics of tumor progression in B-cell malignancies based on Peter Nowell model that tumor progression is multiple step event involving several subsequently occurring cytogenetic changes (18). This is illustrated in Fig.11: during pre B cell development a rare pre B cell develops a t(14;18) translocation involving bcl-2 a low grade B-cell malignancy develops. Then at a certain point during the course of the patients disease a second cytogenetic change occurs in a follicular lymphoma cell carrying a rearranged bcl-2 gene. The 2nd cytogenetic change involves c-myc on chromosome 8. Thus, a cell clone develops that has two translocations, between chromosome 14 and 18, and between chromosome 8 and another chromosome. This cell clone has strong growth advantages and causes the progression of the patients B-cell malignancy into a more malignant phenotype. We had analyzed several cases of acute B-cell leukemia which are characterized by a t(14;18) translocation and additional chromosomal abnormalities involving chromosome 8. In one case, the patient a 5 years history of follicular lymphoma that progressed with a diffuse large cell lymphoma and then developed an acute leukemia with two translocations: a t(14;18) translocation as in follicular lymphoma and an additional t(8;14) translocation as in sporadic Burkitt's lymphoma activating c-myc (19). The other, more interesting case had clinically presented with a very aggressive acute prolymphocytic leukemia which was not responsive to chemiotherapy. Karyotype analysis showed a t(14;18) translocation and multiple additional chromosomal abnormalities including a chr. 17q+. Rearrangement studies showed involvement of c-myc in one of the chromosomal abnormalities, and molecular cloning of the chromosomal breaks revealed that a classical t(14;18) translocation as in follicular lymphoma. More interestingly, we found that the c-myc gene from chromosome 8 had joined into chr.17q22. DNA fragments from the breakpoint region of chromosome 17 were then used as probes and were able to identify a new gene which we called bcl-3. We found bcl-3 highly expressed as 1.7 k.B.transcript in many hematopoietic cell lines of the T-cell, B-cell and myeloid type (Fig.12). Sequence analysis revealed that the bcl-3 promoter had joined into the c-mýc gene in this translocation. We therefore believe that bcl-3 is an activator gene that had activated c-myc in an additional masked t(8;17) translocation resulting in a high grade B cell malignancy, that was derived from a low grade neoplasm with a t(14;18 translocation as in follicular lymphoma (20).

TABLE 1. COMPARISION OF FEATURES OF ENDEMIC AND SPORADIC BURKITT'S LYMPHOMAS.

	Endemic	Sporadic
Geographical Location	Central Africa	Europe and North America
EBV presence	+	-
IgM secretor	-	+
c-myc rearranged	-	+
recombinase-mediated translocation	+	-
isotype switch-mediated translocation	-	+

CONCLUSIONS

What kind of conclusion can we draw from our investigations?

The most human hematopoietic neoplasm carry specific chromosomal abnormalities which are specifically linked to thhe pathogenesis of those disorders.

In B-cell malignancies the scenario of malignant transformation is understood best. In these cases the neoplastic phenotype is the consequence of a reciprocal translocation involving the loci of an activator gene such as the human immunoglobulin heavy chain enhancer or bcl-3 and on the other side a well characterized protooncogene or putative protooncogene. The juxtaposition of an oncogene to such an activator locus results in their transcriptional deregulatioon because of their proximity to genetic elements capable of gene activation in cis also over considerable distances. Sequence analysis of the translocations breakpoints has provided insights with regard to the molecular mechanisms involved in these translocations in B-cells. It appears that the reciprocal translocations in B-cell malignancies are catalyzed by the same enzymes that are involved in physiological immunoglobulin gene rearrangements.

A similar scenario is most likely responsible for the neoplastic transformation in T-cell malignancies. The TCR a/d locus is directly involved in those translocations in T-cell disorders, and is juxtaposed to protooncogene or putative protooncogenes leading to their transcriptional deregulation. But again, the enzyme system involved in the causation of these rearrangements are related to those of B-cell malignancies. Thus, it appears that the molecular basis for T-cell and B-cell malignancies is quite similar; and in the future we can look forward in uncovering the specific role of activator loci and oncogenes in order to draw together a better picture to our understanding of the pathogenesis of the molecular basis of leukemias and lymphomas.

ACKNOWLEDGEMENTS

This work was supported by an Outstanding Investigator Award from the National Cancer Institute CA 39860-05 to Carlo M. Croce, and by a grant from the American Cancer Society to Charlotte E. Gauwerky. We appreciate very much the secretarial assistance of Dawn Branca and Kathy Reinersmann.

REFERENCES

1. Nowell, P.C., Hungerford, D.A., Chromosome of normal and leukemic human leukocytes, Science, 132: 1960; 1497-1500.
2. Yunis, J., The chromosomal basis of neoplasia. Science, 1983; 221: 227-236.
3. Rowley, J.D., Identification of the constant chromosome regions involved in human hematologic malignant disease, Science, 1983; 216: 749-755.
4. Manolov, G., Manolova, Y., Marker band in our chromosome 14 from Burkitt's lymphoma. Nature, 1972; 237: 33-37.
5. Lenoir, G.M. Preud'homaue, G.L., Bernheim, A., Berger, R. Correlation between immunoglobulin light chain expression and variant translocation in Burkitt's lymphoma, Nature, 1982; 298: 474-476.
6. Croce, C.M., Shander, M., Martinis, G., Cicurel, L., D'Ancona, G.G., Dolby, T.W., Koprowsky, H., Chromosomal location of the human immunoglobulin heavy chain genes, Proc. Natl. Acad. Sci., USA, 1981; 76: 3416-3421.
7. Erikson, J., Martinis, G., and Croce, C.M., Assignment of the human genes for immunoglobulin chains to chromosome 22, Nature, 1981, 294: 173-176.
8. McBride, O.W., Heiter, P.A., Hollis, G.F., Swam, D., Otey, M.C., Leder, P., Chromosomal location of human kappa and lambda immunoglobulin light chain constant region genes, J. Exp. Med., 1982, 155: 1480-1491.

9. Malcolm, S., Barton, P., Murphy, C., Ferguson-Smith, M.A., Bentley, D.L., Rabbits, T.H., Localization of human immunoglobulin light chain variable region genes to the short arm of chromosome 2 by in situ hybridization, Proc. Natl. Acad. Sci. , USA, 1982, 79: 4957-4961.
10. Klein, G., The role of gene dosage and genetic transcription in carcinogenesis, Nature, 1981, 294: 313-316.
11. Dalla Favera, R., Bregni, M., Erikson, J., Patterson, D., Gallo, R.C., and Croce, C.M. Human c-myc oncogene is located in the region of chromosome 8 that is translocated in Burkitt's lymphoma cells, Proc. Natl. Acad. Sci., USA, 1982, 79: 7824-7829.
12. Tsujimoto, Y., Cossman, J., Jaffe, E., and Croce, C.M., Involvement of the bcl-2 gene in human folicular lymphoma, Science, 1985, 228: 1440-1443.
13. Tsujimoto, Y., Yunis, J., Ouorato-Showe, L., Erikson, J., Nowell, P.C., and Croce, C.M., Molecular cloning of the chromosomal breakpoint of B-cell lymphomas and leukemias with the t(11;14) chromosome translocation, Science, 1984, 224: 1403-1407.
14. Croce, C.M., Isobe, M., Palumbo, A., Puck, J., Davis, M., and Rovera, G., Gene for a chain of human T-cell receptor: location on chromosome 14 region involved in T-cell neoplasms, Science, 1985, 227: 1044-1047.
15. Chien, Y., Twashima, M., Kaplan, D., Elliot, F., and Davis, M.M., A new T-cell receptor gene located with the alpha locus and expressed early in T-cell differentiation, Nature, 1987, 327: 677-682.
16. Finger, L.R., Harvey, R.C., Moore, R.C.A., Showe, L.C. and Croce, C.M., A common mechanism of chromosomal translocation T- and B-cell neoplasia, Science, 1986, 234: 982-985.
17. Isobe, M., Russo, G., Haluska, F.G., and Croce, C.M., Cloning of the gene encoding the d subunit of the human T-cell receptor reveals its physical organization within the a-subunit locus and its involvement in chromosome translocation in T-cell malignancy, Proc. Natl. Sci., USA, 1985, 85: 3933.
18. Nowell, P.C., The clonal evolution of tumor cell populations, Science, 1976, 194: 23-28.
19. Gauwerky, C.E., Haluska, F.G., Tsujimoto, Y., Nowell, P.C., and Croce, C.M., Evolution of B-cell malignancy: pre B-cell leukemia resulting from MYC activation in a B-cell neoplasm with a rearranged bcl-2 gene, Proc. Natl.acad.Sci., USA, 1988, 85: 8548-8552.
20. Gauwerky, C.E., Huebner, K., Isobe, M., Nowell, P.C., and Croce, C.M., Activation of MYC in a masked t(8;17) translocation results in an aggresive B-cell leukemia, Proc. Natl. Acad. Sci., USA, 1989, 86: 8867-8871.
21. Erikson, J., Finan, J., Nowell, P.C., and Croce, C.M., Translocation of immunoglobulin V_H gene in Burkitt's lymphoma, Proc. Natl. Acad. Sci., USA, 1982, 79: 5611.
22. Croce, C.M., Thierfelder, W., Erikson, J., Nishikura, K., Finan, J., Lenoir, G., and Nowell, P.C., Transcriptional activation of an unrearranged and untranslocated c-myc oncogene by translocation of a Cl locus in Burkitt's lymphoma, Proc. Natl.Acad.Sci., USA, 1983, 80: 6922.
23. Erikson, J., Nishikura, K., ar-Rushdi, A., Finan., J., Emanuel, B., Lenoir, G., Nowell, P.C., and Croce, C.M., Tranlocation of an immunoglobulin locus to a region 3' of an unrearrnged c-myc oncogene enhances c myc transcription, Proc. Natl.Acad.Sci., USA, 1983, 80: 7581.
24. Hayward, W.S. Neel, B.G., Astrin, S.M., Activation of a cellular oncogene by promoter insertions in ALV-induced lymphoid leukosis, Nature, 1981, 290: 465-477.
25. Tonegawa, S. Somati generation of antibody diversity, Nature, 1983, 302: 575-581.
26. Haluska, F.G., Tsujimoto, Y., and Croce, C.M., Mechanisms of chromosome translocation in B- and T-cell neoplasms, Trends Genet, 1987, 3: 11-15.

27. Watt, R., Nishikura, K., Sorrention, J., ar-Rushdi, A., Croce, C.M., and Rovera G., The structure and nucleotide sequence of teh 5' end of the human c-myc gene, Proc.Natl.Acad.Sci., USA, 1983, 80: 6302-6307.
28. Watt ,R., Stanto, L.W., Marcu, K.B., Gallo, R.C., Croce, C.M. and Rovera, G., Nucleotide sequence of cloned cDNA of the human c-myc oncogene, Nature, 1983, 303: 725-728.
29. Nishikura, K., ar-Rushdi, A., Erikson, J., DeJesus, E., Dugan, D., and Croce, C.M., Repression of recombinant m gene and translocated c-myc gene in mouse 3T3 cells, Burkitt's lymphoma cell hybridomas, Science, 1984, 224: 339-403.
30. Siebenlist, U., Hennighause, L., Batbey, J.: Chromatin structure and positive bindin in the putative regulatory region of the c-myc gene in Burkitt's lymphoma, Cell., 1984, 37: 381-391.
31. Croce, C.M., and Nowell, P.C., Molecular basis of human B-cell neoplasia, Blood, 1985; 65: 1-7.
32. Croce, C.M., Tsujimoto, Y., and Erikson, J., Biology of disease: chromosome translocations and B-cell neoplasia, Lab. Invest. 1984, 51: 258-267.
33. Feo, S., Harvey, R., Sharoel, and Croce, C.M., Regulation of translocated c-myc genes transfected into plasmacytoma cells, Proc. Natl. Acad. Sci., USA, 1986, 83: 706-709.
34. Saito, H., Hayday, A.C., Wiman, K., Hayward, W.S., and Tonegawa, S., Activation of the c-myc gene by translocation: a model for translocational control, Proc. Natl. Acad. Sci., USA, 1983, 80: 7476-7480.
35. Ziegler, J.C., Minor, R.C., Rosembaum, E., Lennette, E.T., Shillitoe, E., Casavant, C., Drew, W.L., Muitz, L., Gershow, J., Greenspan, J., Yamanto, K., Outbreak of Burkitt's-like lymphoma in homosexual man, Lancet, , ii, 1982, 631-633.
36. Klein, G., and Klein, E., Evolution of tumor and the impact of molecular oncology, Nature, 1985, 315: 190-195.
37. Haluska, F.G., Russo, G., Kant, G., Andreef, M., and Croce, C.M., Molecular resemblance of an AIDS-associated lymphoma and endemic Burkitt's lymphomas: implications for their pathogenesis, Proc. Natl. Acad. Sci., USA, 1989, 86: 8907-8911.
38. Tsujimoto, Y., Finger, L.R., Yunis, J., Nowell, P.C., and Croce, C.M., Cloning of the chromosomal breakpoint of neoplastic B-cell with the t(14;18) chromosome translocation, Science, 1984, 226: 1097-1099.
39. Pegoraro, L., Palumbo, A., Erikson, J., Fauda, M., Giovanazzo, B., Emanuel, B.S., Rovera, G., Nowell, P.C., and Croce, C.M., a t(14;18) and a t(8;14) chromosomal translocation in a cell line derived from an acute B-cell leukemia, Proc. Natl. Acad. Sci., USA, 1984, 81: 7166-7170.
40. Seto, M., Yaeger, U., Hockett, R.D., Graninger, W., Bennet, S., Goldman, P., Korsmayer, S., Alternative promoters and exons, somatic mutation and deregulation of the bcl-2-Ig fusion gene in lymphoma, The EMBO Journal, Vol. 7, 1988, No. 1, 123-131.
41. Tsujimoto, Y., and Croca, C.M., Analysis of structure, transcripts, and proteins products of bcl-2 the gene involved in follicular lymphoma, Proc. Natl. Acad. Sci., USA, 1986, 83: 5214-5218.
42. Reed, J.C., Tsujimoto, Y., Alpos, J.D., Croce, C.M. and Nowell, P.C., Regulation of bcl-2 oncogene expression during normal human lymphocyte proliferation, Science, 1987, 236: 1295-1299.
43. Tsujimoto, Y., Ikegaki, N., and Croce, C.M., Characterization of the protein product of bcl-2, the gene involved in follicular lymphoma, Oncogene, 1987, 2: 3-7.
44. Haldar, S., Beatty, C., Tsujimoto, Y., and Croce, C.M., The bcl-2 gene encodes a nove g protein, Nature, 1989, 342: 195-198.
45. Tsujimoto, Y., Bashir, M., Gival, I., Cossman, J., Jaffe, E., and Croce, C.M., DNA rearrangements in human follicular lymphoma can involve the 5' or 3' region of the bcl-2 gene, Proc. Natl. Acad. Sci., USA, 1987, 84: 1329-1333.

46. Adachi, M., Cossman, J., Longo, D., Croce, C.M., and Tsujimoto, Y., Variant translocation of the bcl-2 gene to Igl in chronic lymphocytic leukemia, Proc. Natl. Acad. Sci., USA, 1989, 86: 2771.
47. Adachi, M., Teffer, A., Cyreipp, P.R., Kipps, T.J., and Tsujimoto, Y., Preferential linkage of bcl-2 to immunoglobulin light chai gene in chronic lymphocitic leukemia., J. Exp. Med. , 1990, 171: 559-564.
48. Gauwerky, C.E., Jena, N., Catovsky, D., and Croce, C.M., in preparation.
49. Haluska, F.G., Tsujimoto, Y. and Croce, C.M., Oncogene activation in human T-cell and non-Burkitt B-cell malignancies in "Cellular Oncogene Activation", Ed: George Klein, Marcel Dekker, Inc., New York and Basel, 1988, 275.
50. Tsujimoto, Y., Jaffe, E., Cossman, Y., Gorham, Y., Nowell, P.C., and Croce, C.M., Clustering of breakpoints on chromosome 11 in human B-cell neoplasms with the t(11;14) chromosome translocation, Nature, 1985, 315: 340-343.
51. Erikson, J., Finger, L., Jun, L., ar-Rushdi, A., Nishikura, K., Niniowada, J., Finan, J., Emanuel, B., Nowell, P.C., and Croce, C.M., Deregulation of c-myc by translocation of the T-cell receptor in T-cell leukemias, Science, 1986, 232: 884-887.
52. Kagan, J., Finan, J.,, Letofsky, J., Besa, E.C., Nowell, P.C., and Croce, C.M., a-chain locus of the T-cell antigen receptor is involved in the t(10;14) chromosome translocation of T-cell acute leukemia, Proc. Natl. Acad. Sci., USA, 1987, 84: 4543-4546.
53. Erikson, J., Williams, D.L., Finan, J., Nowell, P.C. and Croce, C.M., Locus of the a-chain of the T-cell receptor is split by chromosome translocation in T-cell leukemia, Science, 1985, 229: 784-787.
54. Kagan, J., Finger, L.R., Letofsky, J., Finan, J., Nowell, P.C., and Croce, C.M., Clustering of breakpoints on chromosome 10 in acute T-cell leukemias with the t(10;14) chromosome translocation, Proc. Natl. Acad. Sci., USA, 1989, 86: 4161-4165.
55. Finver, S.N., Martiniere, C., Kagan, J., Cavanee, W., and Croce, C.M., The chromosome 11 region flanking the t(11;14) breakpoint in human T-ALL is deleted in Wilms' tumor hybrids, Oncogene Research, 1989, 5: 143-148.
56. Hecht, F., Morgan, R., Hecht, B., and Smith, S.D., Common region on chromosome 4 in T-cell leukemia and lymphoma, Science, 1984, 226: 1445-1447.
57. Russo, G., Isobe, M., Gatti, R., Finan, J., Batuman, O., Huebner, K., Nowell, P.C., and Croce, C.M., Molecular analysis of a t(14;14) translocation in leukemic T-cells of an ataxia telangiectasia patient, Proc. Natl. Acad. Sci., Usa, 1989, 86: 602-606.
58. Balabon, G.B., Herlyn, M., Clark, W.M., and Nowell, P.C., Karyotypic evolution in human malignant melanoma, Cancer Genet. Cytogenet., 1986, 19: 113-122.
59. Finger, L.W., Kagan, J., Christopher, G., Kurtzberg, J., Involvement of the tcl-5 gene on human chromosome 1 in T-cell leukemia and menanoma, Proc. Natl. Acad. Sci., USA, 1989, 86: 5039-5043.
60. Tsujimoto, Y., Gorham, J., Cossman, J., Jaffe, E., and Croce, C.M., The t(14;18) chromosome translocations involved in B-cell neoplasms result from mistakes in V-D-J joining, Science, 1985, 229: 1390-1393.
61. Haluska, F.G., Finver, S., Tsujimoto, Y., and Croce, C.M., The t(8;14) chromosomal translocation occurring in B-cell malignancies results from mistake in V-D-J joining, Nature, 1986, 324: 158-160.
62. Haluska, F.G., Tsujimoto, Y;, and Croce, C.M., The t (48;14) chromosome translocation of the Burkitt's lymphoma cell line Daudi occurred during immunoglobulin gene rearrangement and involved in the heavy chain diversity region, Proc. Natl. Acad. Sci., USA, 1987, 84: 6835-6939.
63. Desiderio, S.V., Yamcopoulos, G.D., Paskind, M., Thomas, E., Bass, M., Landau, N., Alt, F.W., and Baltimore, D., Insertion of N regions into heavy chain genes is correlated with expression of terminal deoxytransferase in B-cells, Nature, 1984, 311: 752-755.

RECESSIVE MUTATIONS IN HUMAN CANCER PROGRESSION

Webster K. Cavenee

Ludwig Institute for Cancer Research
687 Pine Avenue West
Montreal, Quebec H3A 1A1

INTRODUCTION

Cancer is widely considered to represent the phenotypic manifestation of the accumulation of genetic damage (1); there is a great deal of circumstantial evidence in support of this notion which arises from examination of human populations (2). There has been, for example, extensive documentation of familial aggregation of specific histological types of tumors, sometimes developing with the formal behaviour of an autosomal dominant Mendelian trait. At the level of cytogenetics, various chromosomal aberrations of the germline appear to result in increased propensities for the development of tumors. Tumors often have specific chromosomal rearrangements and, sometimes, such aberrations resemble those which, when inherited, predispose to similar disease (3). The major question which arises from these observations is whether such chromosomal derangements are causal of or caused by the neoplastic process. The challenge is to sort through the bewildering array of such changes and to determine which are common amongst neoplastic diseases at the microscopic and molecular levels and then to categorize them in the context of the large amount of knowledge we have about the process of experimental carcinogenesis.

In any case, the somatic nature of most cancers requires that these genetic abnormalities be acquired during the replication of cells from the specifically affected organ. This would seem to make it likely that they occur during the process of mitotic duplication and segregation of chromosomes from progenitor to daughter cells. Of course, many aberrations in this process would be lethal while others would confer no particular selective advantage. The rare events comprising viable, advantageous and transforming mitotic abnormalities could conceivably represent the molecular underpinnings of the neoplastic process. In this paper, I will review the efforts my colleagues and I have made in the past few years to apply molecular genetics to the question of the specificity of these genetic lesions to the process of cancer predisposition and progression.

GENETICS OF CANCER PREDISPOSITION

Experimental analysis of the process of chemical carcinogenesis has allowed the description of a multistage pathway; the earliest step is termed the initiating

Molecular Basis of Human Cancer
Edited by C. Nicolini, Plenum Press, New York, 1991

event. In the human population such initiations may be transmitted as inherited predisposition. At least fifty different forms of human cancer have been observed to aggregate in families as well as to have corresponding sporadic forms (4). In many of these cases the aggregation occurs with a pattern consistent with the transmission of an autosomal dominant Mendelian trait.

Table I. Loss of Heterozygosity of Chromosome 13q in Retinoblastoma

		Alleles Present at Locus			
Patient	Tissue	D13S1	D13S7	D13S4	D13S5
A. Sporadic:					
RB-LA69	N	**1,2**	1,1	**1,2**	**1,2**
	T	**2**	1	**1**	**2**
RB-409	N	2,2	**1,2**	2,2	1,1
	T	2,2,2	**1,1,1**	2,2,2	1,1,1
RB-412	N	**1,2**	1,1	**1,2**	1,1
	T	**1,2**	1,1	**2,2**	1,1
B. Heritable:					
RB-KS2H	N	2,2	**1,2**	2,2	1,1
	T	2,2	**2,2**	2,2	1,1
RB-462F	N	**1,2***	1,1	**1,2**	**1,2**
	T	**1,2**	1,1	**2,2**	**1,1**

-Not determined; *D13S6 examined, not D13S1; N, Normal; T, Tumor. Alleles designated in bold type are combinations that were heterozygous in constitutional tissue.

One important line of evidence indicating the genetic origin of certain neoplastic diseases is the frequent finding of constitutional chromosomal abnormalities in patients with specific types of tumors. Perhaps the best characterized of these are deletions involving the chromosome region 13q14 which are found in normal tissues of 3-5% of children with bilateral retinoblastoma, a tumor of embryonic neural retina (5). Such deletions may occur de novo, be inherited, or be unbalanced segregants of insertion/deletion translocations. In the latter case, the balanced carriers and family members with duplications do not develop tumors, whereas those with unbalanced deletions of chromosome 13q do (6).

In contrast, tumors that appear as inherited forms often show abnormal segregation of chromosomes, resulting in monosomy and/or duplication of a whole chromosome complement, although structural rearrangements are also frequent. Deletions are mechanistically related to monosomy since both events result in loss of genetic information and due to these events, the tumor cells are hemizygous for part of the genome. Additionally, although retinoblastoma tumor cells often carry

13q-deletions involving chromosome region 13q14, their most frequent (7) rearrangements are structural alterations of chromosomes 1 and 6.

In the context of the requirement for multiple events in tumorigenesis (8), such deletions could act as the first "hit" and, when they are germinal, they could confer the risk of tumor formation in an autosomal dominant manner. Evidence that the same locus is involved in retinoblastoma cases that lack an apparent chromosomal deletion was provided through the demonstration of tight genetic linkage between retinoblastoma and the esterase D locus, the latter being a moderately polymorphic isozymic enzyme whose encoding locus also map to 13q14 (9). This is particularly important since cytogenetically detectable deletions of chromosome band 13q14 are found in only 3-5% of all bilateral retinoblastoma patients and, in most familial form retinoblastoma cases, the primary mutation has not been characterized. However, a model has been proposed (10), which makes specific mechanistic predictions concerning the nature of the chromosomal rearrangements which could serve to unmask the initial predisposing recessive mutation through the somatic attainment of homozygosity for the mutation. This model includes the following chromosomal mechanisms: mitotic recombination between the chromosomal homologues with a breakpoint between the tumor locus and the centromere, which would result in heterozygosity at loci in the proximal region of the chromosome and homozygosity throughout the rest of the chromosome including the tumor locus; mitotic nondisjunction with loss of the wild-type chromosome, either without or with duplication of the mutant chromosome, resulting in hemizygosity or homozygosity at all loci on the chromosome; and, mitotic or regional second events such as gene conversion or point mutation. Sporadic disease could also arise through the appearance of nullizygosity at the tumor locus, the difference being, in this instance, two somatic events as compared to one germinal and one somatic event in heritable cases.

Chromosome specific, single copy segments of the human genome, isolated in recombinant DNA form can be used to recognize polymorphisms at the corresponding chromosomal locus. Sequence variation in restriction enonuclease recognition sites, giving rise to restriction fragment length polymorphism (RFLP) in the locus defined by the probe, are revealed as distinct bands on an autoradiogram and represent alleles of the locus; one from the paternally-derived and one from the maternally-derived chromosomal homologue and behave as Mendelian codominant alleles in family studies. These RFLP markers can be used as linkage markers in inherited disorders, including retinoblastoma. If a disease locus is located close to a polymorphic RFLP marker locus they are likely to segregate together in a family. Therefore, the genotype of DNA markers can be used to infer the genotype at the retinoblastoma locus, and thus to predict if the offspring has inherited the predisposition. Chromosome segregation during tumorigenesis can also be determined in each patient by comparing the child's constitutional and tumor genotypes at each of these marker loci. Recombinant DNA segments have been isolated from human chromosome 13 and used to determine somatic changes in the germline genotypes in several such cases.

Detailed analyses of many retinoblastomas have shown that such events are common and are detected in about 3/4 of all retinoblastoma tumors. These rearrangements fall into four different classes. In 20 of 33 tumors, one constitutional allele was missing at all informative loci along the whole chromosome and 19 of these tumors contained two intact chromosomes 13 as determined either by cytogenetic analysis of the tumor cells or densitometric quantitiations of the autoradiographic signal of the remaining alleles. Therefore, the loss of alleles along the chromosome must involve two separate events; a nondisjunction resulting in loss of one chromosomal complement and either a duplication of the remaining homologue or an abnormal mitotic segregation of the chromosomes resulting in isodisomy as shown for Retin 409 in Table IA. In one case, data consistent with the sole loss of chromosome 13 were obtained. Evidence for mitotic recombination between the chromosome homologues was provided in 4 of the 33 tumors (one example is Retin 412, Table IA). The constitutional genotype was maintained at all informative loci in 9 of the 33 tumors, and therefore, in these cases, the mechanism of attainment of homozygosity could not be determined. These studies strongly suggest that the second event in tumor initiation is comprised of a specific chromosomal rearrangement involving physical loss of the balancing wild type allele at the RB1 locus. This inference was corroborated (11) by examining cases of heritable retinoblastoma and showing that the chromosome 13 homologue retained in these tumors was derived from the affected parent as would be predicted. Two examples are shown in Table IB.

Table II. Loss of Heterozygosity of Chromosome 13q in Osteosarcoma

Patient	Tissue	Alleles Present at Locus			
		D13S1	D13S4	D13S5	D13S3
A. Sporadic:					
OS-03	N	**1,2**	2,2	-	**1,2**
	T	**2,2**	2,2	-	**1,1**
OS-06	N	**1,2**	1,1	**1,2**	1,1
	T	**1,1**	1,1	**1,1**	1,1
OS-09	N	2,2	2,2	2,2	**1,2**
	T	2,2	2,2	2,2	**1,1**
B. Second primary to retinoblastoma:					
OS-1-1	N	**1,2**	**1,2**	**1,2**	2,2
	T	**1,1**	**1,1,2**	**1,2,2**	2,2,2
OS-108	N	**1,2**	2,2	1,1	**1,2**
	T	**1,1,2**	2,2	1,1	**1,1**

-, not determined. N, Normal; T, Tumor. Alleles designated in bold type are combinations that were heterozygous in constitutional tissue.

It is noteworthy that, although the unmasking of predisposing mutations at the RB1 locus occurs in mechanistically similar ways in sporadic and heritable retinoblastoma cases, only the latter carry the initial mutation in each of their cells. Heritable cases also seem to be at greatly increased risk for the development of second primary tumors, particularly osteogenic sarcomas (12). This high propensity may not be merely fortuitous but may be genetically determined by the predisposing RB1 mutation. This notion of a pathogenetic causality in the clinical association between these two rare tumor types was tested (13) by determining the constitutional and osteosarcoma genotypes at RFLP loci on chromosome 13 and representative data are shown in Table II. Osteosarcomas arising in retinoblastoma patients had become specifically homozygous around the chromosomal region carrying the RB1 locus (Table IIB). Furthermore, these same chromosomal mechanisms were observed in sporadic osteosarcomas (Table IIA), suggesting a genetic similarity in pathogenetic causality. These findings are of obvious relevance to the interpretation of human mixed cancer families as they suggest differential expression of a single pleiotropic mutation in the etiology of clinically associated cancers of different histological types.

A likely explanation for the association between retinoblastoma and osteosarcoma is that both tumors arise subsequent to chromosomal mechanisms which unmask recessive mutations in either one common locus that is involved in normal regulation of diferentiation of both tissues, or in separate loci that are located closely within chromosome region 13q14. In either case, germline deletions of the retinoblastoma locus may also affect the osteosarcoma locus. Deletions are likely to be an important form of predisposing mutation at the RB1 locus since a considerable fraction of bilateral retinoblastoma cases carry visible constitutional chromosome deletions and submicroscopic deletions have been detected by reduction of esterase D activity and by molecular analyses using a cDNA for a gene which is, in all likelihood, the transcription product of the retinoblastoma locus (14). The data are also consistent with the chromosomal mechanisms attaining nullizygosity for mutation in a tumor suppressor locus.

The information derived from these studies raises two points relevant to familial predisposition to cancer. Chromosomal mechanisms capable of unmasking predisposing recessive mutations occur in more than one tumor and, at least for chromosome 13, clinically associated tumors share this mechanism of pathogenesis. This latter point suggests that these loci have pleiotropic tissue specificity; however, this pleiotropy appears to be restricted to a small number of tissue types.

GENETICS OF CANCER PROGRESSION

The foregoing section suggests that the attainment of complete defectiveness at a "tumor locus" is one means of initiating the pathway of tumorigenesis. There is no _a priori_ reason that similar mechanisms should be excluded from comprising at least some more distal events as well. In order to test this hypothesis we searched for a disease system characterized by the increasing acquisition of histologically defined malignant criteria. We chose the glial tumors for these first efforts at utilizing genotypic analyses to place tumors into various stages of malignant progression. Gliomas, as a class, are the most common primary neoplasms of the central nervous system. Tumors of this type can be subclassified according to their cellular differentiation, displaying either astrocytic, oligodendrocytic, ependymal, or mixed composition, with astrocytic tumors occurring most frequently. Prognoses for individuals having astrocytoma vary according to the histopathologically assessed malignancy grade of the tumor, however, all adult malignancy grades (grade II-IV) of astrocytoma respond poorly to radiation and/or chemotherapy and the 5-year survival rate for individuals with the most malignant form, glioblastoma (astrocytoma grade IV), is less than 5% (15). The propensity of low malignancy grade astrocytomas to relapse with recurrent tumors that often display a malignant progression accentuates the severity of the disease. Several cytogenetic analyses of high malignancy grade tumors have described frequent chromosome aberrations in direct preparations and short term cultures of astrocytomas (16). In contrast, studies involving astrocytic tumors of low malignancy grade have consistently demonstrated cells with normal karyotypes. This could be due to the analyzed mitoses not being representative of the tumor cells or, alternatively, genetic information could be lost following chromosomal mechanisms not detectable by cytogenetic analysis such as those described in the preceeding section.

In order to determine whether astrocytomas and glioblastomas share a progressional lineage, and whether specific losses of heterozygosity were preferentially associated with some of the stages of the pathway, we initially compared constitutional and tumor genotypes at loci on each human chromosome for samples from 39 adult cases of astrocytoma representing each malignancy grade. Representative data obtained for the loci on chromosome 17p are shown in Table III. This subset of tumors with astrocytic differentiation lost heterozygosity through the loss and duplication or mitotic recombination mechanisms described above, regardless of malignancy stage (17). We cannot now be sure that such events comprise the initial steps; they may occur somewhat later in the progression. They do, however indicate shared insults among all the stages and, as such, in all likelihood represent early events which appear to confer selective advantages used in the outgrowth of the clone. Clearly, it would be desirable to uncover events which occurred at progressively later stages.

Table III. Chromosome 17p in Glioma Stages

Histological Grade	Patient	Tissue	Alleles at Locus		
			GH1	D17S71	D17S5
AII	A3	N	**1,2**	**1,2**	**1,2**
		T	**1,2**	**1,2**	**1,1**
	A5	N	**1,2**	**1,2**	**1,2**
		T	**1,2**	**1,1**	**2,2**
AIII	G2	N	1,1	**1,2**	**1,2**
		T	1,1	**1,2**	**1,1**
	G25	N	**1,2**	**1,2**	**1,2**
		T	**1,2**	**1,1**	**1,1**
AIV	G14	N	**1,2**	**1,2**	**1,2**
		T	**1,1**	**2,2**	**1,1**
	G21	N	**1,2**	**1,2**	**1,2**
		T	**1,2**	**2,2**	**1,1**

-, not determined. Alleles designated in bold type are combinations that were heterozygous in constitutional tissue. N, Normal; T, Tumor.

Cytogenetic analyses have provided important clues to the gross chromosomal changes taking place in these tumors. For example, monosomy for chromosome 10 has been detected in about one third of grade IV astrocytomas (16). We, therefore, compared constitutional and tumor genotypes at loci on chromosome 10 for the 39 cases of astrocytoma analyzed for chromosome 17 above. Allelic combinations were determined with probes homologous to three different chromosome 10 loci: D10S1, D10S4 and PLAU. Representative data obtained with samples of various histological grades are shown in Table IV. Each of 28 grade IV (glioblastoma; GB) tumors examined showed loss of constitutional heterozygosity at one or more of the chromosome 10 loci and these losses appeared to be elicted by nondisjunction resulting in monosomy. In sharp distinction, none of the eleven tumors of lower malignancy grades showed a loss of alleles at any of the chromosome 10 loci examined (18).

Table IV. Chromosome 10 in Glioma Stages

Histological Grade	Patient	Tissue	Alleles at Locus D10S1	D10S4	PLAU
AII	A3	N	1,1	**1,2**	-
		T	1,1	**1,2**	
	A5	N	2,2	**1,2**	**1,2**
		T	2,2	**1,2**	**1,2**
AIII	G2	N	**1,2**	**1,2**	**1,2**
		T	**1,2**	**1,2**	**1,2**
	G25	N	**1,2**	**1,2**	1,1
		T	**1,2**	**1,2**	1,1
AIV	G14	N	**1,2**	**1,2**	**2,2**
		T	**1**	**2**	**2**
	G21	N	1,1	**1,2**	1,1
		T	1	**1**	1

-, not determined. Alleles designated in bold type are combinations that were heterozygous in constitutional tissue. N, Normal; T, Tumor.

These data have two major implications. The first is the demonstration of a clonal origin of the cells comprising these tumors. The cellular pleomorphism of malignant astrocytomas and karyotypic heterogeneity of in vitro derived cell subpopulations arising from primary tumors have complicated attempts to determine the nature of any relationship between the cells which constitute this type of neoplasm. These data show that grade IV astrocytomas arise from the expansion of cells deficient in all or part of chromosomes 10 and 17. The second issue involves the histopathological evidence that astrocytomas progress and become more malignant with time. Because the losses of heterozygosity for chromosome 10 loci were restricted to tumors of the highest malignancy grade, it may be that this aberration is an event of tumor progression, rather than of initiation. Conversely, it is possible that the etiologies and ontogenies of astrocytomas exhibiting low and high degrees of cellular differentiation have no interrelating molecular pathways. However, the postoperative, posttherapeutic recurrence and histological progression of astrocytoma is well documented, providing clinical support for the ontogenic relationship suggested by the shared chromosome 17p aberrations.

CONCLUSIONS

The studies described in this chapter demonstrate the utility of identifying specific genotypic alterations in human tumors. Information so derived bears on each phase of carcinogenesis. When it is used in concert with familial aggregation, inferences concerning predisposition can be drawn and strategies for gene isolation devised. Of course, it is easily envisaged the sorts of applications the information can have viz à viz workforce screening and assignment, actuarial table revision, early disease detection and vigilance in periodic medical examination. When the analyses are applied to a dissection of the pathway of evolution of tumors (1), the opportunities are great as well. For example, one may be able to examine low grade tumors with the object of determining which among them will likely recur or progress. Such information, although only speculative at present would have great medical impact, for example, on therapy selection. All of these possibilities are so tantalizing that the future of genotypic research into the bases of tumor initiation and progression seems bright and exciting indeed.

REFERENCES

1. P.C Nowell, The clonal evolution of tumor cell populations. Science 194: 23 (1976).

2. M. Nordenskjöld and W.K. Cavenee, Genetics and the etiology of solid tumors, in: "Important Advances in Oncology 1988" V.T. De Vita, S. Hellman and S.A. Rosenberg, eds., JB Lippincott, Philadelphia (1988).

3. S. Helm and F. Mitelamn, "Cancer Cytogenetics", Alan R. Liss, New York (1987).

4. J.J. Mulvihill, Genetic repertory of human neoplasia, in: "Genetics of Human Cancer", J.J. Mulvihill, R.W. Miller, J.F. Fraumeni, eds. Raven Press, New York (1977).

5. U. Francke, Retinoblastoma and chromosome 13, Cytogenet. Cell Genet. 16: 131 (1976).

6. L.C. Strong, V.M. Riccardi, R.E. Ferrell, and R.S. Sparkes, Familial retinoblastoma and chromosome 13 deletion transmitted via an insertional translocation, Science 213: 1501 (1981).

7. J. Squire, B.L. Gallie, and R.A. Phillips, A detailed analysis of chromosomal changes in heritable and nonheritable retinoblastoma, Hum. Genet. 70:291 (1985).

8. A.G. Knudson Jr, Mutation and cancer: Statistical study of retinoblastoma, Proc. Natl. Acad. Sci. USA 68: 820 (1971).

9. R.S. Sparkes, A.L. Murphree, R.W. Lingua, M.C. Sparkes, L.L. Field, S.J. Funderburk and W.F. Benedict, Gene for hereditary retinoblastoma assigned to chromosome 13 by linkage to esterase D, Science 219: 971 (1983).

10. W.K. Cavenee, T.P. Dryja, R.A. Phillips, W.F. Benedict, R. Godbout, B.L. Gallie, A.L. Murphree, L.C. Strong and R.L. White, Expression of recessive alleles by chromosomal mechanisms in retinoblastoma, Nature 305: 770 (1983).

11. W.K. Cavenee, M.F. Hansen, E. Kock, M. Nordenskjöld, I. Maumenee, J.A. Squire, R.A. Phillips and B.L. Gallie, Genetic origins of mutations predisposing to retinoblastoma, Science 228:5 501 (1985).

12. F.D. Kitchin and R.M. Ellsworth, Pleiotropic effects of the gene for retinoblastoma, J. Med. Genet. 11: 244 (1974).

13. M.F. Hansen, A. Koufos, B.L. Gallie, R.A. Phillips, O. Fodstad, A. Brogger, T. Gedde-Dahl and W.K. Cavenee, Osteosarcoma and retinoblastoma: a shared chromosomal mechanism revealing recessive predisposition, Proc. Natl. Acad. Sci. USA 82: 6216 (1985).

14. S.H. Friend, R. Bernards, S. Rogelj, R.A. Weinberg, J.M. Rapoport, D.M. Albert, and T.P. Dryja, A human DNA segment with properties of the gene that predisposes to retinoblastoma and osteosarcoma, Nature 323: 643 (1986).

15. P.C. Burger, F.S. Vogel, S.B. Green and T.A. Strike, Glioblastoma multiforme and anaplastic astrocytoma. Pathologic criteria and prognostic implications, Cancer 56: 1106 (1985).

16. S.H. Bigner, J. Mark, M.S. Mahaley, and D.D. Bigner, Patterns of the early gross chromosomal changes in malignant human gliomas, Hereditas 101: 103 (1989).

17. C.D. James, E. Carlbom, M. Nordenskjöld, V.P. Collins and W.K. Cavenee, Mitotic recombination of chromosome 17 in astrocytomas, Proc. Natl. Acad. Sci. USA 86: 2858 (1989).

18. C.D. James, E. Carlbom, J.P. Dumanski, M. Hansen, M. Nordenskjöld, V.P. Collins and W.K. Cavenee, Clonal genomic alterations in glioma malignancy stages, Cancer Res. 48: 5546 (1988).

TRANSFORMING GROWTH FACTOR-β REGULATION OF EPITHELIAL PROLIFERATION

Jennifer A. Pietenpol, Edmund Y. Yang and Harold L. Moses

Department of Cell Biology, Vanderbilt University School of Medicine, Nashville, Tennessee, 37232, USA

ABSTRACT

The closely related mammalian TGFßs (TGFß1, TGFß2 and TGFß3) are potent inhibitors of proliferation of many cell types in vitro. TGFß1 has also been demonstrated to be growth inhibitory in vivo for epithelial, endothelial, myeloid and lymphoid cells. Utilizing skin keratinocytes as a model system for studying the mechanism of TGFß1-induced growth inhibition, it has been demonstrated that TGFß1 rapidly inhibits transcription of the c-myc gene. Antisense c-myc oligonucleotides were found to inhibit proliferation of the keratinocytes as effectively as TGFß1, indicating that TGFß1 suppression of c-myc expression is an important component in the mechanism of TGFß growth inhibition. Studies utilizing DNA tumor virus transforming gene constructs have provided evidence for the necessity of the retinoblastoma gene product, pRb, or a related protein, in the pathway for TGFß1 suppression of c-myc transcription. Thus, TGFß1 may act through a tumor suppressor gene product, pRb, to negatively regulate transcription of a protooncogene, c-myc, and subsequently inhibit cell proliferation.

There have been remarkable advances in our understanding of the diffusible factors that regulate cell proliferation and the molecular mechanisms involved in this regulation over the past decade. Most normal cells and tissues exist within a complex environment of growth stimulatory and growth inhibitory factors. Mechanisms of endocrine, paracrine and autocrine growth regulation are likely to play a very important role in normal cellular processes including proliferation, development, and differentiation[2-3]. Alteration in these normal cellular processes may result from a modification in the balance between growth stimulatory and growth inhibitory signals. To data the majority of work has focused on the actions of the growth stimulatory peptides and their relationship to cellular protooncogenes. It is now widely recognized that protooncogenes code for growth factors, growth factor receptors, molecules involved in growth factor signal transduction, and transcription factors that are necessary for the growth factor response (reviewed in Goustin et al.[15]). Work with oncogenes has been particularly important in identifying genes important in the growth factor-receptor-response pathway and in pinpointing important regulatory steps in this growth stimulatory response.

Molecular Basis of Human Cancer
Edited by C. Nicolini, Plenum Press, New York, 1991

While negative regulators of cell proliferation have been hypothesized to be important for a number of decades, this field did not gain respectability until the demonstration that TGFß1 is a potent inhibitor of proliferation of many cell types[41].TGFß1 remains as the prototypic diffusible negative regulator of cell proliferation and is known to have a remarkable diversity of biological effects in addition to the inhibition of cell proliferation (reviewed in Lyons and Moses[21]). TGFß1 is now known to belong to a very large family of related molecules with a wide range of regulatory activities in embryologic development and in the adult state (reviewed in Miller et al[25]). Of the numerous TGFß-like molecules, the mammalian TGFß1, TGFß2 and TGFß3 proteins have been shown to be potent inhibitors of cell proliferation[16]. Recent studies from our laboratory[31] have indicated that antioncogenes, or tumor suppressor genes, are probably involved in the response pathway for the negative regulation of cell proliferation by TGFß1.

TGFβ1 PROCESSING AND ACTIVATION

TGFβ1 is a dimeric polypeptide comprised of identical 112-amino-acid subunits whose association is maintained through disulfide linkages[1]. Cloning and sequencing of the TGFß1 gene revealed that TGFß1 is synthesized as a large molecule of 390 amino acids[9]. The pre-pro TGFß1 contains a typical hydrophobic signal sequence of 29 amino acids which is cleaved resulting in pro-TGFß1 as depicted in Figure 1. Several glycosylation sites have been identified within the amino-terminal portion of the TGFß1 precursor[13] and are hypothesized to play an important role in processing and secretion[37]. At some point during synthesis or transit, disulfide bond formation and dimerization occurs. While the ordering of these processing steps and the tertiary structure of the TGFß1 molecule represented are unknown, it has become apparent that the synthesis of TGFß1 is complex involving several processing events[12-13-21-22-23-26-27]. As shown in Figure 1, proteolytic cleavage at a dibasic cleavage site (residue 278) separates the mature 25 kDa TGFß1 from the amino-terminal glycopeptide. These polypeptides remain non-covalently associated from the latent TGFß1 complex.

Processed TGFß1 is secreted from most cells in culture and released from platelets in the inactive or latent form (Lawrence et al., 1984 and Pircher et al., 1986). TGFß1 in the latent form cannot interact with cell surface receptors and therefore cannot elicit biological activities. The activation of TGFß1 is likely to be a key regulatory step in TGFß action. Acidification or alkalization of cell-conditioned medium, or acid extraction of tissue causes disruption of non-covalent interactions between the 25 kDa mature TGFß1 and the amino-terminal glycopeptide portion of the TGFß1 precursor (Figure 1)[22]. While acidification is capable of disruption of these noncovalent interactions, thus releasing active TGFß1, protease activation has been achieved as well. Specifically, plasmin, a serine protease, appears to activate latent TGFß1 by cleavage within the amino-terminal glycopeptide[21-22-23]. This may result in conformation changes which destabilize the latent complex and ultimately release mature TGFß1 (Figure 1). Experiments by Sato and Rifkin[35-36], support proteolytic cleavage of TGFß1 by plasmin as a relevant physiological mechanism of activation. They observed that the in vivo activation of TGFß1, during co-culture of bovine endothelial cells and pericytes, could be blocked by inhibitors of plasmin action.

REGULATION OF CELL PROLIFERATION BY TGFβs

Although originally described as a factor that stimulated growth in soft agar of rodent fibroblasts[28-33], it is unlikely that TGFß1, or any of the TGFßs, are direct mitogens for any cell type. They are probably indirect mitogens for certain

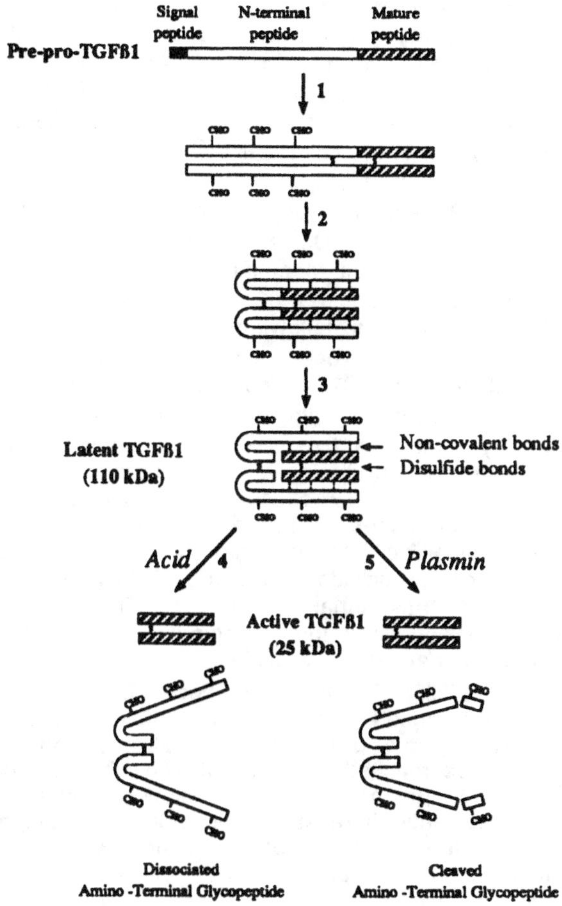

FIG.1 Model of sunthesis and activation of latent TGFβ1. Proposed processing events of the TGFβ1 precursor (1-3). Disulfide bond formation and dimerization occur at some point during synthesis and transit . The order of the these processing steps , and the tertiary structure of the TGFβ1 molecule represented are unknown. The 25-kDa TGFβ1 homodimer (crosshatched box) remains noncovalently associated with amino-termianl glycopeptide (open box) after proteolytic cleavage at the dibasic cleavage site (3). Events which disrupt this complex, such as changes in ionic strenght (4) or alterations in conformation (5), may be destabilize the latent complex and allow the release of mature, biuologically active TGFβ1.

mesenchymal cells through induction of autocrine stimulation by endogenous growth factors[20]. TGFß1 was first demonstrated to be a growth inhibitor in studies comparing the biological activities of the growth inhibitor from African green monkey (BSC-1) cells originally described by Holley et al[18] and human platelet derived TGFß1[41]. It is now known that the BSC-1 growth inhibitor is identical to the TGFß2 gene[17]. Subsequent studies have demonstrated that TGFß1 and TGFß2 are the most potent growth inhibitory polypeptides known for a wide variety of cell types in culture including most epithelial cells, endothelial cells, most lymphoid cells, and many myeloid cells (for review, see Bascom et al[3]). TGFß3 has also been shown to be growth inhibitory for epithelial cell types in culture[16-25]. In general, the inhibitory effects of the TGFßs are reversible, and where studied, the TGFßs cause growth arrest in the G_1 phase of the cell cycle[38-6].

TGFß1 and TGFß2 have also been demonstrated to be inhibitory for cell proliferation in vivo. Silberstein and Daniel[39] reported that TGFß1 administered in semi-solid pellets reversibly inhibits mammary epithelial cell proliferation in vivo. Russell et al[34] found that TGFß1 and TGFß2, administered intravenously, inhibited the early phase of liver regeneration following partial hepatectomy. Inhibition of hematopoietic cell proliferation has been demonstrated by Goey et al[14] following intraarterial administration of TGFß1. Evidence has been presented for TGFß1 suppression of the immune response to immunogenic tumors in vivo[40].

Recent studies by Yang and Moses[43] have demonstrated that TGFß1 has remarkable effects on the chicken chorioallantoic membrane (CAM). TGFß1 induced gross angiogenesis and, histologically, caused the rapid induction of a hypercellular lesion characterized by increases in the density of epithelial, fibroblastic and endothelial cells. Substantial increases in fibroblast and epithelial cell numbers were observed as early as 4 hr following administration of TGFß1 which was imbedded in a methylcellulose disk placed on the surface of the CAM. This increase in cell number in the central portion of the lesion where the concentrations of TGFß1 would be expected to be highest was accompanied by a marked inhibition of cell proliferation as demonstrated by incorporation of ^{3}H-thymidine detected by autoradiography. Because cells increased in number rapidly in the face of inhibition of cell proliferation, the effects were almost certainly due to stimulation of cell migration through chemotactic effects. Once the cells had reached the areas of higher TGFß1 concentrations, their proliferation was inhibited. Inhibition of proliferation of epithelial, fibroblastic and endothelial cells was observed. There was some evidence of compensatory increased cellular proliferation at the periphery of the lesion where exogenous TGFß1 concentrations would be expected to be very low or non-existent.

These studies are reminiscent of in vitro studies where it has been demonstrated that the chemotactic effects of TGFß1 occur at a much lower (sub-picomolar) concentration than that required for other TGFß1 effects[32]. At higher concentrations (picomolar levels) TGFß1 is no longer chemotactic and other biologic effects are induced, including growth inhibition and simulation of extracellular matrix production. We speculate that TGFß1 administered to the CAM causes a rapid inward migration of cells toward the higher concentrations of TGFß1 through the process of chemotaxis. Once cells have reached the zone of higher TGFß1 concentration, cell proliferation is inhibited so that cells may devote their energy to performing differentiated function which in the case of the fibroblasts is deposition of extracellular matrix and in the case of the endothelial cells is formation of capillary tubes and secretion of basement membrane materials. These studies may have important implications for the mechanisms of TGFß action in the promotion of wound healing. They further demonstrate that TGFß1 is inhibitory in this in vivo model for epithelial cells, fibroblasts, and endothelial cells[43].

TGFβs EFFECTS ON EARLY GROWTH FACTOR INDUCED EVENTS

In studying the mechanisms of TGFß growth inhibition, we have utilized secondary cultures of human skin keratinocytes grown in low-calcium, serum-free medium[29-38] and a continuous line of mouse keratinocytes cultured in low-calcium, serum-containing medium[5-6]. Both cell types require EGF/TGF_ for proliferation and retain the ability to differentiate under high calcium conditions. TGFß1, TGFß2 and TGFß3 are potent inhibitors for the keratinocytes[6-16]. The cells were found to be reversibly inhibited in their growth by the TGFßs with the majority of the cells blocked in the G_1 phase of the cell cycle. Half maximal inhibition was obtained at approximately 10 pM. The keratinocytes also synthesized and released TGFß1 into the conditioned medium with confluent cultures producing as much as 80 pM in 24 hr[38-6]. However, all of the TGFß1 released was in a latent form detectable only after acid treatment of the conditioned medium, similar to observations with many other cell types. Whether this latent TGFß1 activates spontaneously or can be activated by the cells with subsequent binding of the active TGFß to cell surface membrane receptors is not known. However, since keratinocytes have receptors for the TGFßs, are capable of responding to the factors, and secrete relatively large quantities into the medium, the possibility of negative autocrine regulation by TGFß1 in these cells must be entertained as a viable possibility.

In considering the mechanisms by which the TGFßs inhibit cell proliferation, it is known that the growth inhibitory effect is not secondary to cytotoxicity; the inhibition is reversible[38-6]. Induction of terminal differentiation, although reported for bronchial epithelial cells[24], does not appear to be a general phenomenon and does not occur in either the human foreskin keratinocytes or in the mouse keratinocyte cell line[38-6]. It is possible that prior growth inhibition of keratinocytes by the TGFßs may render them more susceptible to induction of differentiation by physiologically inducers such as calcium. TGFß1 does not appear to interfere with growth factor-receptor interactions or with many of the early events induced by growth stimulatory factors including some signal transduction events[4-6]. For example, EGF induction of c-fos in mouse keratinocytes is not inhibited by TGFß1 even though EGF stimulation of DNA synthesis is inhibited[5].

Perhaps the best evidence that TGFß does not interfere with early growth factor induced events is the data demonstrating that TGFß1 can prevent EGF-stimulated DNA synthesis in the mouse keratinocytes even when added late in G_1, long after early events induced by EGF have occurred[31]. Following EGF addition to the mouse keratinocytes that have been growth arrested by depletion of EGF, there is a 12 hr lag phase prior to entry of the cells into the S phase of the cell cycle[6]. TGFß1 was shown to inhibit subsequent entry into S phase if added at any time up to the G_1/S boundary[31]. Thus, in considering mechanisms of TGFß growth inhibition, events other than those induced early by growth factors must be examined.

THE ROLE OF c-myc IN GROWTH INHIBITION BY TGFβ

c-myc has been shown to be necessary for cell proliferation in a number of systems (for review, see Bascom et al[3]. Like the other immediate-early genes, it is induced rapidly by the addition of growth factors to quiescent cells of many types including the mouse keratinocytes[5]. However, unlike the other immediate early genes, c-myc expression remains elevated throughout G_1 and during S phase[5] (J.A. Pietenpol,

R.M. Lyons, J.T. Holt, and H.L. Moses, unpublished work, 1990). The addition of TGFß1 to rapidly growing human and mouse keratinocytes resulted in a rapid reduction of c-myc mRNA and protein[5]. Half-maximal reduction in c-myc mRNA following TGFß addition occurred in approximately 30 min. Since the half-life of c-myc mRNA following addition of actinomycin D is approximately 20 min, the effects of TGFß on suppression of c-myc must occur quite rapidly, yet protein synthesis is probably required for TGFß suppression of c-myc[5]. In quiescent mouse keratinocytes stimulated with EGF, TGFß1 has been demonstrated to rapidly reduce c-myc when added at any point during G_1 or the early part of S phase (J.A. Pietenpol, R.M. Lyons, J.T. Holt, and H.L. Moses, unpublished work, 1990). Since TGFß rapidly reduces c-myc, we have investigated whether c-myc is necessary for cell proliferation in the mouse keratinocytes. The addition of antisense c-myc oligonucleotides was demonstrated to rapidly reduce c-myc protein and to inhibit cell proliferation as effectively as TGFß1[30]. This suggests that TGFß suppression of c-myc is important in the mechanism of TGFß1 induced growth inhibition.

The mechanism of TGFß suppression of c-myc was investigated. Nuclear run-on assays utilizing single-stranded, exon- and intron-specific probes demonstrated that the block in c-myc expression by TGFß1 occurred at the level of transcriptional initiation[30]. Studies with the series of 5' deletion c-myc/chloramphenicol acetyltransferase expression vectors indicated that a cis regulatory element which resides between positions -100 and +71 relative to P1 transcription start site, is necessary for the TGFß1 suppression. Based on this data, it is proposed that the mechanism of TGFß1 growth inhibition involves synthesis or modification of a protein that may interact with a specific cis element in the 5' regulatory region of c-myc gene resulting in inhibition of transcriptional initiation of this gene[30]

pRb IS NECESSARY FOR TGFβ1 SUPPRESSION OF c-myc TRANSCRIPTION

In considering other proteins that may be involved in TGFß1 regulation of c-myc transcription, it became apparent that the protein product of the retinoblastoma gene, pRb, was a likely candidate. pRb has properties of a cell cycle regulatory factor[8]. It is a differentially phosphorylated 105 kd nuclear protein that has been postulated to contribute to growth suppression, directly or indirectly[11] .Evidence of tumor suppressor activity by pRb is provided by the observations that loss or inactivation of the Rb gene has been found in a variety of malignant cell types including retinoblastomas, osteosarcomas, small cell carcinoma of the lung, and carcinomas of the breast, bladder and prostate (see Pietenpol et al[31] for review). Furthermore, introduction of a cloned Rb gene into cells results in marked changes in morphology, decreased growth rates, and reduced tumorigenic potential in neoplastic cells[19].

Additional evidence for a role of pRb in growth regulation has come from studies with DNA tumor virus transforming gene products. It has been demonstrated that pRb associates with the transforming proteins of SV40[7], adenovirus[42] and HPV-16[10]. The specific transforming proteins are large T antigen (TAg), E1A and E7, respectively. Furthermore, mutant forms of these virus proteins that do not bind pRb are transformation defective. In the case of TAg, binding to pRb only occurs when pRb is in a hypophosphorylated state in the G_1 phase of the cell cycle[8]. It has been hypothesized that binding of pRb by DNA viral transforming proteins may block the growth suppressive activity of pRb resulting in some of the aberrations of growth control associated with transformation by these viruses[7-42].

In recent studies keratinocytes transformed by the DNA tumor viruses SV40 and HPV-16 or HPV-18 were shown to be resistant to the growth inhibitory effects of TGFß1[31]. Further, TGFß1 was unable to suppress c-myc mRNA levels in these DNA tumor virus-transformed cells. Therefore, we tested whether blocking of TGFß1 suppression of c-myc expression is mediated through the pRb binding domain of the

DNA tumor virus transforming proteins. Transient expression in non-transformed keratinocytes of the DNA tumor virus proteins that bind pRb including HPV-16 E7, adenovirus E1A and SV40 TAg, prevented TGFß1 suppression of c-myc expression. Mutants of E1A and TAg that impaired association with pRb failed to block the TGFß1 suppression of c-myc. The observed blocking effect of the DNA tumor virus transforming proteins did not appear to be a general consequence of tumor virus transforming function because an E1A mutant that is unable to transform but remains able to bind pRb continues to block TGFß1 suppression of c-myc transcription. These observations suggest that TGFß1-induced growth inhibition is a consequence of pRb-mediated repression (directly or indirectly) of c-myc gene transcription. The DNA tumor virus transforming proteins that bind to pRb probably render pRb inactive in this pathway.

However, proteins other than pRb bind to the pRb binding domain of the DNA viral transforming proteins[42], and their involvement in this inhibitory pathway cannot be completely excluded at this time.

SUMMARY

The TGFßs are potent inhibitors of proliferation for most cell types in vitro and in vivo. The mechanism of growth inhibition by TGFßs has been investigated utilizing cultures of skin keratinocytes. In quiescent cells stimulated with epidermal growth factor (EGF), TGFß1 can inhibit DNA synthesis when added at any point in the G_1 phase of the cell cycle up to the G_1/S boundary. TGFß1 rapidly reduced the amount of c-myc mRNA and protein by inhibiting transcriptional initiation. Like TGFß1, antisense c-myc oligonucleotides inhibited c-myc protein synthesis and cell proliferation. SV40 and human papilloma virus (HPV) transformed human keratinocytes were shown to be resistant to the growth inhibitory and c-myc suppression effects of TGFß1. Since the DNA tumor viruses have in common the binding and inactivation of the retinoblastoma tumor susceptibility gene product (pRb), it was hypothesized that pRb was an intermediate in the pathway for TGFß1 suppression of c-myc and growth inhibition. Transient expression of HPV-16 E7, adenovirus E1A, and SV40 T antigen blocked TGFß1 suppression of c-myc transcription. This effect was not observed with DNA tumor virus transforming proteins mutated in their pRb binding domains such that they no longer bind pRb. These and other observations suggest that pRb or a protein that is able to bind to the DNA tumor virus transforming proteins mediates TGFß1 regulation of c-myc gene expression and growth inhibition.

REFERENCES

1. Assoian, R.K., Komoriya, A., Meyers, C.A., Miller, D.M., and Sporn, M.B., 1983, Transforming growth factor-beta in human platelets: Identification of a major storage site, purification, and characterization. J. Biol. Chem. 258:7155-7160.
2. Barnard, J.A., Bascom, C.C., Lyons, R.M., Sipes, N.J., and Moses, H.L., 1988, Transforming growth factor ß in the control of epidermal proliferation. Am. J. Med. Sci. 31:159-163.
3. Bascom, C.C., Sipes, N.J., Coffey, R.J., and Moses, H.L., 1989, Regulation of epithelial cell proliferation by transforming growth factors. J. Cell. Biochem. 39:25-32.
4. Chambard, J.-C. and Pouysségur, J., 1988, TGF-ß inhibits growth factor-induced DNA synthesis in hamster fibroblasts without affecting the early mitogenic events. J. Cell. Physiol. 135:101-107.
5. Coffey, R.J., Bascom, C.C., Sipes, N.J., Graves-Deal, R., Weissman, B.E., and Moses, H.L., 1988a, Selective inhibition of growth-related gene expression in murine keratinocytes by transforming growth factor ß. Mol. Cell. Biol. 8:3088-3093.

6. Coffey, R.J., Sipes, N.J., Bascom, C.C., Graves-Deal, R., Pennington, C.Y., Weissman, B.E., and Moses, H.L., 1988b, Growth modulation of mouse keratinocytes by transforming growth factors. Cancer Res. 48:1596-1602.
7. DeCaprio, J.A., Ludlow, J.W., Figge, J., Shew, J.-Y., Huang, C.-M., Lee, W.-H., Marsilio, E., Paucha, E., and Livingston, D.M., 1988, SV40 large tumor antigen forms a specific complex with the product of the retinoblastoma susceptibility gene. Cell 54:275-283.
8. DeCaprio, J.A., Ludlow, J.W., Lynch, D., Furukawa, Y., Griffin, J., Piwnica-Worms, H., Huang, C.M., and Livingston, D.M., 1989, The product of the retinoblastoma susceptibility gene has properties of a cell cycle regulatory element. Cell 58:1085-1095.
9. Derynck, R., Jarrett, J.A., Chen, E.Y., Eaton, D.H., Bell, J.R., Assoian, R.K., Roberts, A.B., Sporn, M.B., and Goeddel, D.V., 1985, Human transforming growth factor-beta complementary DNA sequence and expression in normal and transformed cells. Nature 316:701-705.
10. Dyson, N., Howley, P.N., Münger, K., and Harlow, E., 1989, The human papilloma virus-16 E7 oncoprotein is able to bind to the retinoblastoma gene product. Science 243:643-646.
11. Friend, S.H., Bernards, R., Rogelj, S., Weinberg, R.A., Rapaport, J.M., Albert, D.M., and Dryja, T.P., 1986, A human DNA segment with properties of the gene that predisposes to retinoblastoma and osteosarcoma. Nature 323:643-646.
12. Gentry , L.E., Webb, N.R., Lim, G.J., Brunner, A.M., Ranchalis, J.E., Twardzik, D.R., Lioubin, M.N., Marquardt, H., and Purchio, A.F., 1987, Type 1 transforming growth factor beta: Amplified expression and secretion of mature and precursor polypeptides in chinese hamster ovary cells. Mol. Cell. Biol 7:3418-3427.
13. Gentry, L.E., Lioubin, M.N., Purchio, A.F., and Marquardt, H., 1988, Molecular events in the processing of recombinant type 1 pre-pro-transforming growth factor beta to the mature polypeptide. Mol. Cell. Biol. 8:4162-4168.
14. Goey, H., Keller, J.R., Back, T., Longo, D.L., Ruscetti, F.W., and Wiltrout, R.H., 1989, Inhibition of early murine hemopoietic progenitor cell proliferation after in vivo locoregional administration of transforming growth factor-ß1. J. Immunol. 143:877-880.
15. Goustin, A.S., Leof, E.B., Shipley, G.D., and Moses, H.L., 1986, Growth factors and cancer. Cancer Res. 46:1015-1029.
16. Graycar, J.L., Miller, D.A., Arrick, B.A., Lyons, R.M., Moses, H.L., and Derynck, R., 1989, Human transforming growth factor-ß3: Recombinant expression, purification, and biological activities in comparison with transforming growth factors-ß1 and -ß2. Mol. Endocrinol. 3:1977-1986.
17. Hanks, S.K., Armour, R., Baldwin, J.H., Maldonado, F., Spiess, J., and Holley, R.W., 1988, Amino acid sequence of the BSC-1 cell growth inhibitor (polyergin) deduced from the nucleotide sequence of the cDNA. Proc. Natl. Acad. Sci. USA 85:79-82.
18. Holley, R.W., Armour, R., and Baldwin, J.H., 1978, Density-dependent regulation of growth of BSC-1 cells in cell culture: Growth inhibitors formed by the cells. Proc. Natl. Acad. Sci. USA 75:1864-1866.
19. Huang, H.-J.S., Yee, J.-K., Shew, J.-Y., Chen, P.-L., Bookstein, R., Friendmann, T., Lee, E.Y.-H., and Lee, W.-H., 1988, Suppression of the neoplastic phenotype by replacement of the RB gene in human cancer cells. Science 242:1563-1566.
20. Leof, E.B., Proper, J.A., Goustin, A.S., Shipley, G.D., DiCorleto, P.E., and Moses, H.L., 1986, Induction of c-sis mRNA and activity similar to platelet-derived growth factor by transforming growth factor-ß: A proposed model for indirect mitogenesis involving autocrine activity. Proc. Natl. Acad. Sci. USA 83:2453-2457.
21. Lyons, R.M. and Moses, H.L., 1990, Transforming growth factors and the regulation of cell proliferation. Eur. J. Biochem. 187:467-473.

22. Lyons, R.M., Keski-Oja, J., and Moses, H.L., 1988, Proteolytic activation of latent transforming growth factor-ß from fibroblast conditioned medium. J. Cell Biol. 106:1659-1665.

23. Lyons, R.M., Gentry, L.E., Purchio, A.F., and Moses, H.L., 1990, Mechanism of activation of latent recombinant transforming growth factor ß1 by plasmin. J. Cell Biol. 110:1361-1367.

24. Masui, T., Wakefield, L.M., Lechner, J.F., LaVeck, M.A., Sporn, M.B., and Harris, C.C., 1986, Type ß transforming growth factor is the primary differentiation-inducing serum factor for normal human bronchial epithelial cells. Proc. Natl. Acad. Sci. USA 83:2438-2442.

25. Miller, D.A., Pelton, R.W., Derynck, R., and Moses, H.L., 1990, Transforming growth factor ß: A family of growth regulatory peptides. Ann. N.Y. Acad. Sci., in press.

26. Miyazono, K., Hellman, U., Wernstedt, C., and Heldin, C.H., 1988, Latent high molecular weight complex of transforming growth factor beta-1: Purification from human platelets and structural characterization. J. Biol. Chem. 263:6407-6415.

27. Miyazono, K. and Heldin, C.H., 1989, Role of carbohydrate structures in TGF-beta1 latency. Nature 338:158-160.

28. Moses, H.L., Branum, E.B., Proper, J.A., and Robinson, R.A., 1981, Transforming growth factor production by chemically transformed cells. Cancer Res. 41:2842-2848.

29. Moses, H.L., Tucker, R.F., Leof, E.B., Coffey, R.J., Halper, J., and Shipley, G.D., 1985, Type ß transforming growth factor is a growth stimulator and a growth inhibitor. In: J Feramisco, B Ozanne, and C Stiles (eds.), Growth Factors and Transformation. Cancer Cells 3, pp. 65-71. Cold Spring Harbor, NY: Cold Spring Harbor Press.

30. Pietenpol, J.A., Holt, J.T., Stein, R.W., and Moses, H.L., 1990a, Transforming growth factor ß1 suppression of c-myc gene transcription: Role in inhibition of keratinocyte proliferation. Proc. Natl. Acad. Sci. USA 87:3758-3762.

31. Pietenpol, J.A., Stein, R.W., Moran, E., Yaciuk, P., Schlegel, R., Lyons, R.M., Pittelkow, M.R., Münger, K., Howley, P.M., and Moses, H.L., 1990b, TGFß1 inhibition of c-myc transcription and growth in keratinocytes is abrogated by viral transforming proteins with pRB binding domains. Cell, 61:777-785.

32. Postlethwaite, A.E., Keski-Oja, J., Moses, H.L., and Kang, A.H., 1987, Stimulation of the chemotactic migration of human fibroblasts by transforming growth factor ß. J. Exp. Med. 165:251-256.

33. Roberts, A.B., Anzano, M.A., Lamb, L.C., Smith, J.M., and Sporn, M.B., 1981, New class of transforming growth factors potentiated by epidermal growth factor: Isolation from non-neoplastic tissues. Proc. Natl. Acad. Sci. USA 78:5339-5343.

34. Russell, W.E., Coffey, R.J., Ouellette, A.J., and Moses, H.L., 1988, Type ß transforming growth factor reversibly inhibits the early proliferative response to partial hepatectomy in the rat. Proc. Natl. Acad. Sci. USA 85:5126-5130.

35. Sato, Y. and Rifkin, D.B., 1989, Inhibition of endothelial cell movement by pericytes and smooth muscle cells: Activation of a latent transforming growth factor ß1-like molecule by plasmin during co-culture. J. Cell Biol. 109:309-315.

36. Sato, Y., Tsuboi, R., Lyons, R.M., Moses, H.L., and Rifkin, D.R., 1990, Characterization of the activation of latent TGFß by co-culture of endothelial cells and pericytes or smooth muscle cells: A self-regulating system. J. Cell Biol., in press.

37. Sha, X., Brunner, A.M., Purchio, A.F., and Gentry, L.E., 1989, Transforming growth factor beta: Importance of glycosylation and acidic proteases for processing and secretion. Mol. Endocrinol. 3:1090-1098.

38. Shipley, G.D., Pittelkow, M.R., Wille, J.J., Scott, R.E., and Moses, H.L., 1986, Reversible inhibition of normal human prokeratinocyte proliferation by type ß

transforming growth factor-growth inhibitor in serum-free medium. Cancer Res. 46:2068-2071.
39. Silberstein, G.B. and Daniel, C.W., 1987, Reversible inhibition of mammary gland growth by transforming growth factor-ß. Science 237:291-293.
40. Torre-Amione, B., Beauchamp, R.D., Koeppen, H., Park, B.H., Schreiber, H., Moses, H.L., and Rowley, D.A., 1990, A highly immunogenic tumor transfected with a murine transforming growth factor type ß1 cDNA escapes immune surveillance. Proc. Natl. Acad. Sci. USA. 87:1486-1490.
41. Tucker, R.F., Shipley, G.D., Moses, H.L., and Holley, R.W., 1984, Growth inhibitor from BSC-1 cells closely related to the platelet type ß transforming growth factor. Science 226:705-707.
42. Whyte, P., Buchkovich, K.J., Horowitz, J.M., Friend, S.H., Raybuck, M., Weinberg, R.A., and Harlow, E., 1988, Association between an oncogene and an anti-oncogene: the adenovirus E1A proteins bind to the retinoblastoma gene product. Nature 334:124-129.
43. Yang, E.Y. and Moses, H.L., 1990, Transforming growth factor ß1-induced changes in cell migration, proliferation, and angiogenesis in the chicken chorioallantoic membrane. J. Cell Biol., in press.

b-N-ACETYLHEXOSAMINIDASE IN LEUKAEMIC CELLS

Carla Emiliani, Sabata Martino, John L. Stirling° and Aldo Orlacchio

Dipartimento di Medicina Sperimentale e Scienze Biochimiche Università di Perugia, 06100 Perugia, Italy, and ° M.R.C. Human Genetic Diseases Research Group, Department of Biochemistry, King's College London, Kensington Campus, Campden Hill, London W8 7AH, U.K.

In normal tissues the lysosomal enzyme b-N-acetylhexosaminidase (E.C.3.2.1.52) is expressed as two major forms, A and B, with the subunit composition ab and bb respectively. The two subunits of human b-N-acetylhexosaminidase, a and b, are encoded by separate genes located on chromosomes 15 and 5, respectively. The genes are 35-40 kilobases long and have similar architecture; 12 of the 13 introns interrupt the coding regions at analogue positions[18-19]. As is a characteristic of proteins destined for lysosomes, the two subunits of b-N-acetylhexosaminidase are transported through the endoplasmic reticulum and Golgi. They undergo numerous post-traslational modifications of mature subunits therefore differ significantly from those of the newly synthetized polypeptides[14]. A post traslation event of particular importance to b-N-acetylhexosaminidase is the dimerization of its subunits to give catalitically active enzyme. The a- and b- monomers can associate in three ways to give homo- or heterodimeric isoenzymes: A (ab) and S (aa). The a- and b- subunits are synthesized in cultured cells in approximately equal amounts[10] but they dimerize at different rates. The three isoenzymes have a different isoelectric point, molecular weight and thermostability. They can be easily separated by conventionaml ionic exchange chromatographies or by biospecific affinity separation techniques[20-6].
The A (ab) isoenzyme has the broadest specificity. Whereas both A and B isoenzymes can remove nonreducing terminal G1cNAc and Ga1NAc residues from all glycopeptides, glycosaminoglycans and glycolipids that occur in human cell and artificial substrates, only b-hexosaminidase A is effective towards synthetic substrates in which the sugar is 6-sulphated[12-1]. This ability is conferred by the presence of active a-subunit. The S (aa) isoenzyme has generally limited catalytic activity and is unstable. Considerable research interest has focussed on the biochemical genetics of human hexosaminidase because of the ganglioside storage diseases which result from the inactivity of hexosaminidase A. If the A isoenzyme activity is lost because of mutation in either the a- or b- subunit gene, it cannot be replaced by the action of either B or S isoenzyme, and GM_2 ganglioside accumulates in lysosomes.
Glycosphingolipids (GSLs) are ubiquitous membrane components and have been shown to be located almost exclusively at the outer leaflet of plasma membranes[11]. Dramatic changes in GSL composition and metabolism have been observed during differentiation and oncogenic transformation, suggesting a specific role of membrane GSLs in the regulation of cell growth and cellular interaction[9].

Molecular Basis of Human Cancer
Edited by C. Nicolini, Plenum Press, New York, 1991

Nojiri et al.[15] demonstrated that, when human myeloid and monocytic leukaemic cell lines HL 60 and U 937, respectively, were treated with an exogenous sialoglycosphingolipid, ganglioside GM_3, cell growth was markedly inhibited, and their morphological maturation along a monocytic lineage was observed, indicating that ganglioside molecular species may play an important role in the growth and differentiation of leukaemic cell lines.

HL-60 is a human promyelocytic leukemia cell line[4] widely used for analysing cellular and molecular events during the induction of myeloid differentiation and we were interested to find that it exhibits drammatic alterations in b-N-acetilhexosaminidase activity and isoenzyme expression[7]. In our characterization of b-N-acetilhexosaminidase in HL-60 we found an increased specific activity and a major change in isoenzyme pattern, with the B form almost lacking and the appearance of a new "extra" isoenzyme, which represented 40% of total activity, and resulted to be very similar to b-N-acetylhexosaminidase S (aa), which is the residual activity in patients with Sandhoff's disease.

In order to evaluate whether expression of the "extra" form is related to cell proliferation we have examined changes in b-N-acetylhexosaminidase activity and isoenzymes during the growth cycle of HL-60. Modification or modulation of b-N-acetylhexosaminidase activity may have some part in changing ganglioside content of leukemic cell membrane and as a consequence may be in some way related to the regulaion of cell growth and differentiation.

MATERIALS AND METHODS

Culture conditions

The human promyelocytic leukaemia cell line HL-60 and fetal human fibroblasts were grown in RPMI-1640 medium (GIBCO Europe, Breda, Netherlands) supplemented with 10% heat-inactivated fetal calf serum (GIBCO Europe) in a humidified atmosphere containing 5% CO_2.

For experimental cultures, HL-60 cells were taken from cultures at a density of 2.0 x 10^6/ml, resuspended in fresh culture medium at 0.25 x 10^6/ml and dispensed (10 ml) into culture flasks. At each indicated time, one flask was harvested, the cells were counted and cell viability was estimated and was generally in excess of 95%. The cells were then separated from the medium by centrifugation and both cells and medium were collected.

Confluent fibroblasts were treated with trypsine, then centrifuged, resuspended in fresh culture medium and dispensed into 50 mm Petri dishes at the cell concentration of 0.25 x 10^6 cells/dish. At each indicated time, one dish was harvested, the cells counted and viability estimated as above.

Preparation of cell lysates and enzyme extraction

As described by Emiliani et al.[7], cell samples were suspended in 1 ml of 10 mM-sodium phosphate buffer, pH 6.0 and 0.1% (v/v) Miranol H2M, an ionically balanced detergent, after three washings with 0.9% naCl, then sonicated and finally centrifuged at 36,000 g (r_{av} 8.19 cm) for 20 min. The supernatants were used as cell lysates.

Enzyme assay

Activities of the enzymes b-N-acetihexosaminidase an a-D-mannosidase were estimated using the appropriate fluorogenic substrates 4-methylumbelliferyl-b-N-acetylglucosaminide (4-MUG1cNAc) or 4-methylumbellyferyl-a-D-mannoside (4-MU-a-mann) (all from Koch-Light Lab., Haverhill, Suffolk, U.K.). The activity of b-hexosaminidase alpha subunits was estimated using the specific substrate 4-

methylumbelliferyl-b-N-acetylglucosamine-6-sulphate ($4MUG1cNAc-6-SO_4$), obtained as an impure preparation from Koch-Light Lab. and purified by the Beccari et al. (1987) method. All the substrates were at a concentration of 3 mM, except $4MUG1cNAc-6-SO_4$ which was used at 0.3 mM. The substrates 4MUG1cNAc and $4MUG1cNAc-6-SO_4$ were dissolved in 0.1 M citric acid/0.2 M phosphate buffer, pH 4.5 whereas 4MU-a-mann was dissolved in the same buffer but at pH 4.0.
One enzyme unit (U) was defined as the amount of enzyme that hydrolyzed 1 mmol of substrate/min at 37°C.
Protein content was determined by the Bradford[3] method using crystalline bovine gamma globulin as a standard.
Specific activities were expressed as enzyme units/mg of protein or enzyme units/10^6 cells.

DEAE-cellulose chromatography

The chromatography was performed according to the method used by Robinson and Stirling[20], using a 1 ml column equilibrated with 10 mM-Na phosphate buffer, pH6.0. Enzyme activity retained by the column was eluted in a linear gradient of NaCl that reached a concentration of 0.5 M in 50 ml of buffer. The column was then washed with 1 M-NaCl in column buffer. Fractions (0.5 ml) were collected and assayed for activity towards the substrates 4-MUG1c-NAc, $4-MUG1cNAc-6-SO_4$ and 4MU-a-mann.

RESULTS

Changes in the total intracellular and extracellular activities of b-N-acetylhexosaminidase during the growth of HL-60

During the growth of HL-60 (Fig. 1, A) the total intracellular activity of b-N-acetylhexosaminidase showed a number of remarkable changes. These were detected with both 4-MUG1cNAc and $MUG1cNAc-6-SO_4$, although the changes were more exaggerated when the latter substrate was used. When HL-60 cells, collected at a density of 2 x 10^6 cells/ml, were resuspended in fresh medium at a density of 2.5 x 105 there was a rapid and sustained decrease in the activity per cell which reached its lowest point after 6 to 8 h. Hexosaminidase specific activities decreased in this time from 16.6 x10^{-3} U/mg of protein to 2.5 x 10^{-3} u/mg of protein when assayed with 4-MUG1cNAc.
In marked contrast to this behaviour, HL-60 cells resuspended in fresh medium at the same density as they were when harvested (2 x 10è cells/ml) showed no decrease of b-N-acetylhexosaminidase activity.

After 12 h the cells began to grow exponentially and there was a gradual increase in the activity per cell which was sustained until 72 h. Thereafter the rate of accumulation of increased and maximal activities per cell were reached after 120 h. At this stage in growth the specific activities were 15 x 10^{-3} and 2 x 10^{-3} U/mg of protein for the activities measured with $4-MUG1cNAc-6-SO_4$ respectively.
b-N-acetylhexosaminidase activity accumulated in the medium throughout growth, but the rate of accumulation was greatest between 96 and 120 h (Fig. 1, B). It was noticeable that the rapid loss of intracellular activity was not accompanied by a corresponding increase in extracellular activity.
b-N-acetylhexosaminidase was not the only lysosomal glycosidase to show changes in specific activity during the growth of HL-60. Results for a-D-mannosidase are shown in Fig. 1, A and B, and closely follow the gross changes observed with b-hexosaminidase.

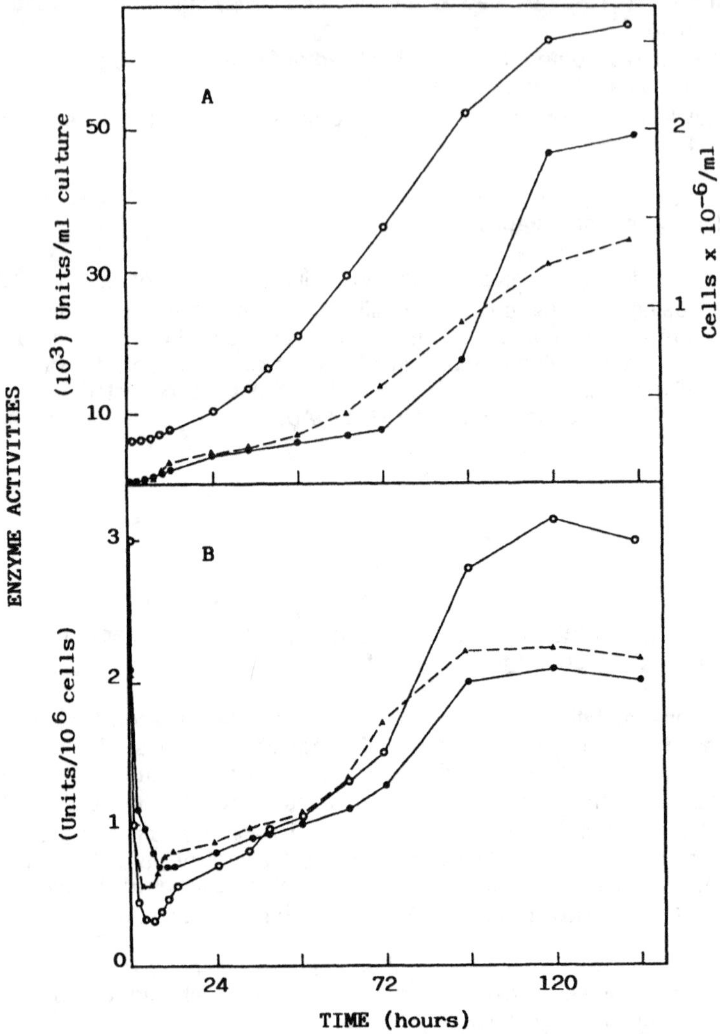

FIG. 1. Changes in total b-N-acetylhexosaminidase and a-D-mannosidase in cells and medium during growth of HL 60 cells.

A: m, growth curve of HL 60 cells; l, (10^3) mmol 4MUG1cNAc hydrolyzed x min^{-1}; s, (10^3) mmol 4MU-a-mann hydrolyzed x min $^{-1}$.

B: l, (10^3) mmol AMUG1cNAc hydrolyzed x min $^{-1}$; s, (10^3) mmol 4MU-a-mann hydrolyzed x min $^{-1}$; m, (10^4) mmol 4MUG1cNAc-6-SO_4 hydrolyzed x min^{-1}.

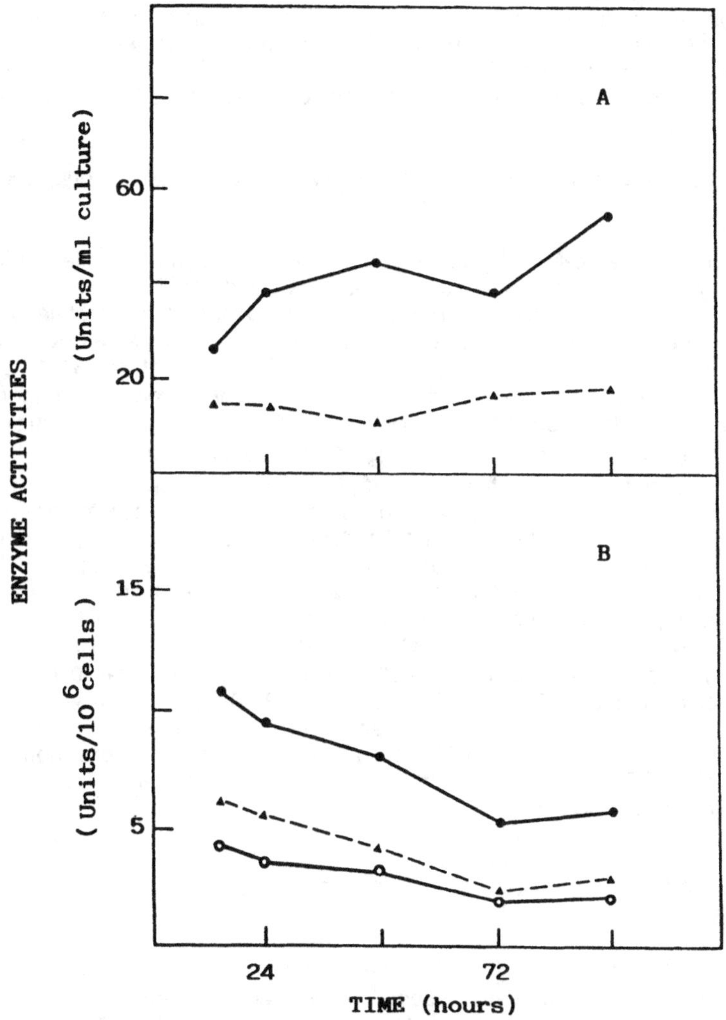

FIG. 2. Intracellular and extracellular total b-N-acetylhexosaminidase and a-D-mannosidase in fetal fibroblasts at different stages of growth.

A:Activities secreted in the medium; l, (10^3) mmol 4MUG1cNAc hydrolyzed x min $^{-1}$; s, (10^5) mmol 4MU-a-mann hydrolyzed x min $^{-1}$.

B:Intracellular activities. l, (10^3) mmol 4MUG1cNAc hydrolyzed x min^{-1}; s, (10^3) mmol 4MU-a-mann hydrolyzed x min^{-1}; m, (10^4) mmol 4MUG1cNAc-6-SO_4 hydrolyzed x min $^{-1}$.

Changes in the total extracellular and intracellular activity of b-N-acetylhexosaminidae in normal fibroblasts during growth

Normal human fetal fibroblast secrete b-N-acetylhexosaminidase in the medium (Fig. 2,A). An accumulation of enzyme activity in the medium was detected when cells became confluents, with the highest value after 96 hours in culture. The variations detected in extracellular enzyme activity could be justified by a different secretion of enzyme in the medim since a decrease in enzyme activity per cell corresponded to the increased activity in medium (Fig. 2, B).
Another lysosomal enzyme, a-D-mannosidase, behaved similarly (Fig. 2, A-B).

Changes in intracellular b-N-acetylhexosaminidase isoenzymes during the growth of HL-60

HL-60 has a b-N-acetylhexosaminidase isoenzyme profile which differs considerably from that of normal tissues and leukaemic cell lines. Very small amounts (less than 1% of total activity) of hexosaminidase B, the form unretained by DEAE-cellulose, are present in HL-60 and the acidic A form is accompanied by an even more acidic "extra" form. The "extra" form has a higher ratio of MUG1cNAc-6-SO_4/4-MUG1cNAc activity than the A form, (Fig. 3, C). Underlying the changes in total hexosaminidase activity during the growth of HL-60 are alterations to the isoenzyme profiles. In the experiment shown in Fig. 3, HL-60 cells (2 x 10^6 cells/ml) were collected, resuspended in fresh medium (2,5 x 10^5) and allowed to grow for up to 96 h. Extracts of cells collected at 6, 48 and 96 hours were analyzed for hexosaminidase isoenzymes by chromatography on DEAE-cellulose. Three main changes were apparent a reduction in the activities under the A and "extra" peaks, an elution of the "extra" peak at a lower salt concentration indicating a shift to a less acidic net charge and an increased ability of the "extra" form to hydrolyze the sulphated substrate, as indicated by the ratio MUG1cNAc-6-SO_4/4-MUGcNAc, that was 0.35 and 0.40 at 6 h and 48 h respectively whereas in high density cells (96 h in culture) it was 0.17. All changes were further accentuated in cells after 48 h (Fig. 3, B), then there was a substantial restoration of the b-N-acetylhexosaminidase A and "extra" peaks. A number of satellite peaks were present (Figs.3, A-B) which were very much more noticeable with MUG1cNAc-6-SO_4 as the substrate. The elution position and sizes of these wee remarkably reproducible.

Intacellular b-N-acetylhexosaminidase isoenzymes in confluent and not-confluent fibroblasts

As illustrated by Fig. 4, fetal fibroblasts have an isoenzyme profile which is very similar to that of normal human tissues[13] or normal leukocytes[7] with a B isoenzyme (bb) representing about 30% of total activity, A (ab) 60% and very little intermediate form, entirely composed of b-subunit, as revealed by sulphated substrate analysis. Also a-D-mannosidase isoenzyme pattern behave similarly to that of normal human tissues and leukocytes.
No significant differences in isoenzymes patterns of either b-N-acetylhexosaminidase or a-D-mannosidase were detected by analysing not-confluent (Fig. 4, A) or confluent fibroblasts (Fig. 4, B).

DISCUSSION

It is now evident that there are a number of alterations to the total activity and isoenzymes in leukaemic cells[8-2-5] but as yet the origins of these changes are unknown. We have previously described a novel form of b-N-acetylhexosaminidase in fresh populations from patient with myeloid and lymphoid leukemia[17] This form

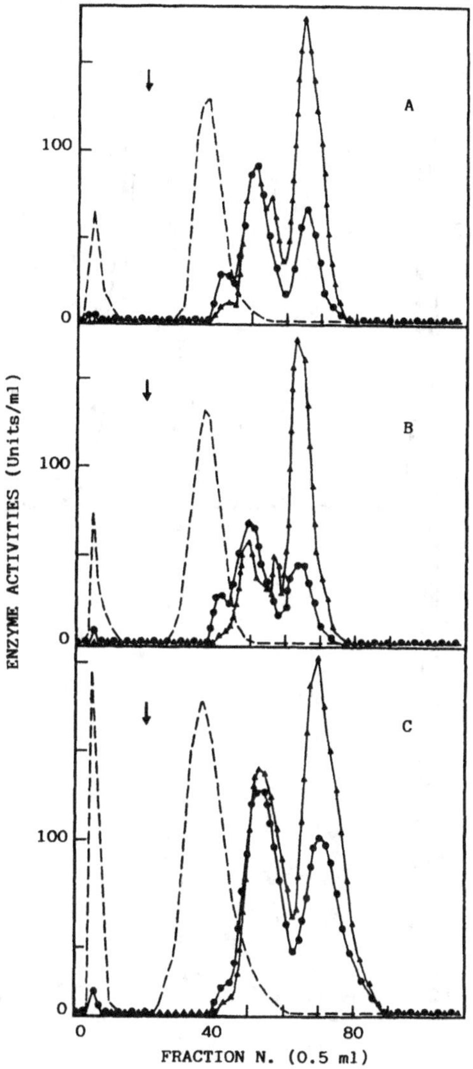

FIG. 3. b-N-acetylhexosaminidase and a-D-mannosidase in HL-60 cells at different stages of growth.

Extracts of 10^7 cells harvested at A:6h; B:48h; C:96h were chromatographed on a 1 ml DEAE-52 column equilibrated with 10 mM Na-phosphate buffer, pH 6.0. Enzyme activity retained by the column was eluted in a linear gradient of NaCl (arrow) that reached a concentration of 0.5 M in 50 ml of buffer. Fractions were collected and assayed for enzyme activities.

Units are expressed as: l, (10^5) mmol 4MUG1cNAc hydrolyzed x min^{-1}; ---, (10^5) mmol 4MU-a-mann hydrolyzed x min^{-1}; s, (10^6) mmol 4MUG1cNAc-6-SO_4 hydrolyzed x min^{-1}.

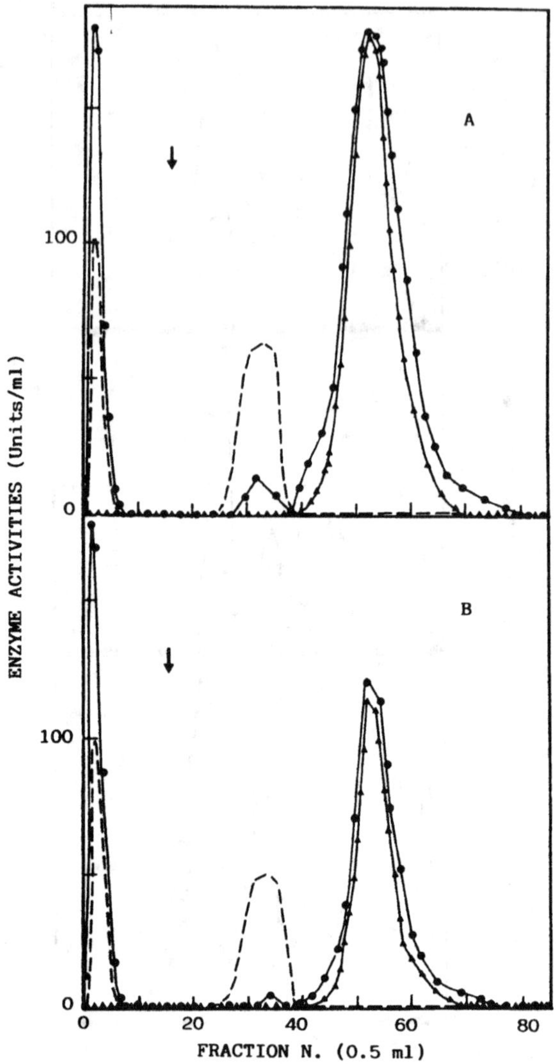

FIG. 4. b-N-acetylhexosaminidase and a-D-mannosidase isoenzymes in human fetal fibroblast at different stages of growth.

Extracts of 4 x 10^6 cells harvested at A:24 h (cells were not-confluent) and B:96 h (confluent cells) were chromatographed as described in Fig. 3.

Units are expressed as: l, (10^5) mmol 4MUG1cNAc hydrolyzed x min^{-1}; --, (10^7) mmol 4MU-a-mann hydrolyzed x min^{-1}; s, (10^6) mmol 4MUG1cNAc-6-SO_4 hydrolyzed x min^{-1}.

No "extra" isoenzymes of a-D-mannosidase were detected in HL 60 cells (Fig. 3, C) compared to the normal fibroblasts or to the normal tissues and leukocytes.

Unlike b-hexosaminidase there were no alterations to the elution position of a-mannosidase isoenzymes in the salt gradient when cells were analyzed at different times during growth (Fig. 3, A-C).

which we identified as "extra" form, is also expressed in a number of leukemic cell lines and we were interested to find that this form represented 40% of total activity in the promyelocytic cell line HL 60. In the same cells, the B form, that normally represents 30% of total activity, was almost lacking. HL 60 also exhibited an increased specific activity of b-N-acetylhexosaminidase, particularly evident when measured using the sulphated substrate, specific for active a-subunit. The presence of a-subunit confer to b-N-acetylhexosaminidase A the stability to hydolyze the ganglioside GM_2 and a further interest in the expression of b-N-acetylhexosaminidase in HL 60 cells is drawn on the observation that gangliosides added exogenously to cells in culture can alter their developmental fate[9-15]. In our characterization of the extra form of b-N-acetylhexosaminidase in HL 60 we concluded that it had properties very similar to those of b-N-acetylhexosaminidase S, the form representing the residual activity in Sandhoff's disease, the more extreme case of b-subunit limitation. An hypothesis of mutations in HL 60 cells inactivating the b-subunit locus is very unlikely because some b-subunit is however expressed and normally aggregates to give b-N-acetylhexosaminidase A. Our data are more consistent with a regulated expression of a- or b-subunits. b-N-acetylhexosaminidase is an ubiquitous enzyme but its expression could however be modified in abnormal cells like leukaemic cells. We wondered whether the "extra" form was expressed throughout growth or was a form characteristic of one of the growth phases. Subculturing HL 60 cells caused a rapid decrease in the specific activity of b-N-acetylhexosaminidase, that was not due to losses incurred during centrifugation or washing prior to suspension in fresh medium. The decrease was a progressive one which took place over 6-8 hours and contrasted with the behaviour of lactate dehydrogenase, myeloperoxidase, unspecific granulocyte collagenase, lysozyme and chymotrypsin-like cationic protein which maintained its specific activity throughout the growth of HL 60 cells[16]. By contrast when fetal fibroblasts were analyzed for intracellular and extracellular b-N-acetylhexosaminidase during growth there was no evidence of loss of activity in the early phases of growth. The decrease in intracellular activity in the latest phases of growth could be explained by an increased secretion in the culture medium. Whatever the explanaton for the initial decline in b-hexosaminidase in HL 60 cells, it would also seem to apply to the a-D-mannosidase. One possibility is the loss of a sub-class of lysosomes on transfer to fresh medium accompanied by rapid degradation of the released enzymes by proteinases.

Both fibroblasts and HL 60 cells were analyzed for their isoenzyme content during growth. No changes were recorded in the isoenzyme patterns of fibroblasts analyzed at different cell densities. Different isoenzyme profiles, with an elution of the "extra" peak at a lower salt concentration, indicating a shift to a less acidic net charge, were evidentiated in HL 60 cells. In marked contrast to this behaviour HL 60 cells resuspended in fresh medium but at high density, as they were when harvested, showed no decrease of b-N-acetylhexosaminidase activity and no changes in isoenzyme profile, suggesting that the structural changes observed during HL 60 growth could be strictly related to cell density. The biochemical basis of this is not known but the shift to a less acidic net charge would be consistent with the preferential degradation of sub-forms of the isoenzyme carrying a higher concentration of negatively charged substituents such as phosphate, sulphate or sialic acids.

DEAE-cellulose chromatography also evidentiated an increased ability of the "extra" form to hydrolyze $4MUG1cNAc\text{-}6\text{-}SO_4$, the substrate which is hydrolyzed by a-subunits and not by b-subunits. At present it is not known if a different specificity towards the natural substate GM_2 ganglioside corresponds to this increased ability to hydrolyze the synthetic substrate. Whatever the explanation, no changes would be seen to apply to the B isoenzyme (which does not contain a-subunits) or to other lysosomal enzymes like a-mannosidase which mantain the same isoenzyme pattern at all cell densities investigated.

b-N-acetylhexosaminidase modifications that take place during the growth of HL 60 cells particularly envolve the a-subunit. Since the a-subunit is the only cause of the hydrolysis of the GM_2 ganglioside, b-N-acetylhexosaminidase may have some part in changing the ganglioside content of cell membrane and as a consequence may be in some way related to cell growth regulation and differentiation.

ACKNOWLEDGEMENTS

Work supported by Italian C.N.R. and M.P.I.

REFERENCES

1. Beccari, T., Emiliani, C., Hosseini, R., Orlacchio, A. and Stirling, J.L., 1987 "Intermediate forms of b-N-acetylhexosaminidase lack activity towards 4-methylumbelliferyl-b-N-acetylglucosaminide-6-sulphate" Biochem. J. 244: 801.
2. Besley, G.T.N., Moss, S.E., Bain, A.D., Dewar, A.E., 1983 "Correlation of lysosomal enzyme abnormalities in various forms of adult leukaemia" J. Clin. Pathol., 36: 1000.
3. Bradford, M.M., 1976 "A rapid and sensitive method for the quantitation of microgram quantities of protein utilizing the principle of protein-dye binding" Anal. Biochem. 72: 248.
4. Collins, S.J., Gallo, R., Gallagher, R., 1977 "Continuous growth and differentiation of human myeloid leukaemic cells in suspension culture" Nature, 270: 347.
5. Drexler, A.G., Gaedicke, G., Novotny, J., Minowada, J., 1986 "Occurrence of particular isoenzymes in fresh and cultured leukaemia-lymphoma cells" Cancer, 58: 251.
6. Emiliani, C., Beccari, T., Stirling, J.L., Sciarra, R., Orlacchio, A., 1988 "New biospecific chromatographies for b-hexosaminidase" in "Macromolecular Biorecognition", I. Chaiken, E. Chiancone, A. Fontana and P. Neri Eds., The Humana Press Inc., Clifton, N. J., USA.
7. Emiliani, C., Beccari, T., Tabilio, A., Orlacchio, A., Hosseini, R., Stirling, J.L., 1990 "An enzyme with properties similar to those of b-N-acetylhexosaminidase S is expressed in the promyelocytic cell line HL 60" Biochem. J., 267: 111.
8. Ellis, R.B., Rapson, N.T., Patrik, A.D., Greaves, M.F., 1978 "Expression of hexosaminidase isoenzymes in childhood leukemia" N. Engl. J. Med., 268: 476.
9. Hakomori, S., 1981 "Glycosphingolipids in cellular interaction, differentiation and oncogenesis" Ann. Rev. Biochem., 50: 733.
10. Hasilik, A. and Neufeld, E.F., 1980 "Biosynthesis of lysosomal enzymes in fibroblasts" J. Biol. Chem., 255: 4937.
11. Karlsson, K.A., 1976 "Aspects on structure and function of sphingolipids in cell surface membranes" in "Structure of biological membranes" Abrahamsson S., Pascher, I. Eds., Plenum New York.
12. Kytzia, H.J., Sandhoff, K., 1985 "Evidence for two different active sites on b-hexosaminidase A" J. Biol. Chem., 260: 7568.
13. Mahuran, D., Novak, A., Lowden, J.A., 1985 "The lysosomal hexosaminidase isoenzymes" in "Isoenzymes: current topics in biological and medical research", Vol.12, New York, Alan R. Liss Inc.
14. Neufeld, E.F., 1989 "Natural history and inherited disorders of a lysosomal enzyme, b-hexosaminidase", J. Biol. Chem., 264: 10927.
15. Nojiri, H., Takaku, F., Terui, Y., Miura, Y., Saito, M., 1986 "Ganglioside GM3: an acidic membrane component that increases during macrophage-like cell differentiation of human myeloid and monocytoid leukemic cell lines HL 60 and U 937" Proc. Natl. Acad. Sci. USA, 83: 782.
16. Olsson, I. and Olofsson, T., 1981 "Induction of differentiation in a human promyelocytic leukemic cell line (HL 60)" , Exp. Cell. Res. , 131:225.
17. Orlacchio, A., Maffei, C., Emiliani, C., Rambotti, P., Davis, S., 1984 "A distinct b-hexosaminidase isoenzyme separated from human leukemic lymphocytes and myelocytes" , Biochem. Biophys. Res. Commun., 122: 966.
18. Proia, R.L. and Soravia, E., 1987 "Organization of the gene encoding the human b-hexosaminidase a-chain", J. Biol. Chem., 262: 5677.

19. Proia, R.L., 1988 "Gene encoding the human b-hexosaminidase b-chain: extensive homology of intron placement in the a- and b-chain genes", Proc. Natl.Acad. Sci., USA, 85: 1883.
20. Robinson, D. and Stirling , J.L., 1968 "N-acetyl-b-glucosaminidases in human spleen", Biochem. J., 107: 321.

THE ROLE OF SMALL t ANTIGEN IN SV40 ONCOGENESIS

M. Carbone, F. Pompetti, C. Cicala, P. Nguyen, K. Dixon, and A. S. Levine

Section on Viruses and Cellular Biology, NICHD
Bethesda MD 20982
USA

INTRODUCTION

Simian virus 40 (SV 40) is an oncogenic papovavirus capable of inducing tumors in rodents and of transforming cells from many species in tissue culture (Topp. et al. 1981). Due to the small size of its genome, about 5000 bp, SV40 has been extensively studied as a model for understanding mechanism of viral oncogenesis in rodents.

The viral proteins associated with in vivo oncogenesis and in vitro cell transformation are encoded by the early region of the virus; these proteins are known as the T (tumor) antigens because animals bearing tumors induced by SV40 have antibodies against these viral proteins. Two SV40 tumor antigens are detected in infected or transformed cells, "large T" and "small t". These two proteins are produced from a single viral gene by differential splicing of an RNA transcript (Topp et al. 1981). Unlike the oncogenes of the retroviruses, the T-antigen oncogenes of SV40, and those of popaviruses in general, do not have a homologus cellular protooncogene (for a review see Carbone and Levine 1990).

The transforming capacity of SV40 and the different roles played by the two SV40 oncogenic proteins have been studied both in in vitro and in vivo systems.

IN VITRO STUDIES

The large T antigen

SV40 large T antigen is a 96 kDa polypeptide found mainly in the nuclei of infected and transformed cells. The large T antigen is the principal gene product responsible for cell transformation by SV40 and it is required for cell transformation under all conditions tested (Topp et al. 1981). For many years the mechanism by which T antigen transformed cells remained enigmatic. Large T can induce cellular DNA replication but this effect does not in itself seem to account for its ability to act as an oncogenic protein (Butel 1986). Induction or modification of cellular transcription factors is an attractive mechanisl for altering cellular gene expression, and it might contribute to cellular transformation. However, there is no direct evidence for such a

Molecular Basis of Human Cancer
Edited by C. Nicolini, Plenum Press, New York, 1991

mechanism in the case of SV40 (Butel 1986). Recently it has been suggested that SV40 large T antigen promotes transformation because of its known interactions with several cellular proteins (for a review, see Green 1989). This hypothesis has been supported by the discovery that two of the cellular proteins that physically interact with the large T antigen correspond to the products of the cellular tumor suppressor genes p53 and RB (for a review, see Carbone and Levine 1990, Lane and Benchimol 1990). These two cellular gene products are both necessary to prevent the cell from cycling and in the normal cell, they must be inactivated through phosphorylation/dephosphorilation events to permit the resting cells to traverse from G1 to S (Bishoff et al. 1990, Cooper and Whyte 1989). Papovaviruses, including SV40, normally infect differentiated cells that are growth arrested. These quiescent cells must be stimulated to enter S phase for efficient viral DNA replication to occur. By complexing with the p53 and RB proteins, the large T antigen may inactivate these cellular proteins and permit the cell to enter the S phase. Thus, T antigen facilitates the division of infected cells, enabling efficient viral replication (and cell lysis) in the normal host.

Transformation may occur when the virus infects a cell that is unable to support viral replication, and T antigen induces continuous cycling. Interestingly, several lines of evidence suggest that inactivation of the p53 and/or RB gene products predisposes a normal cell to become transformed. For example, in humans, the appearance of several tumors such as retinoblastoma, osteosarcoma, and colon carcinoma seems to be associated with the inactivation of RB and p53 (for a review, see Fearon and Vogelstein 1990, Weinberg 1989).

In the past, although useful for basic studies, the SV40-induced hamster tumor model was considered to have only limited relevance to tumor development in humans. However, the recent discovery of associations between viral oncoproteins and cellular tumor suppressor gene products suggests that DNA tumor viruses remain excellent models with which to illuminate molecular mechanism related to human carcinogenesis.

The small t antigen

In contrast to the vast body of literature on the role of large T antigen in SV40 transformation, the function of SV40 small t antigen remains elusive. The small t antigen is a 17-21 kDa protein found predominantly in hte cytoplasm of infected and transformed cells (Topp et al. 1981). This protein shares 82 amino acids at its amino terminus with the large T antigen, while the remaining 92 amino acids are unique. By comparing small t deletion mutants with the wild-type (wt) virus, several functions have been assigned to small t. Small t mutants fail to transform resting cells: Once the cells are in growing state, however, these mutants transform the cells at a normal frequency (Martin et al. 1979; Seif and Martin, 1979; Sleigh et al. 1978). In addition, small t is able to overcome the growth-arresting effects of theophylline in CVI cells (Rundell and Cox, 1979). Small t also interacts with two cellular proteins, recently identified (Pallas et al. 1990, Walter et al. 1990) as the catalytic and the regulatory subunits of phosphatase 2A (pp2A). It is tempting to speculate that this interaction might influence the state of phosphorylation of T antigen, which has been showm to be a substrate for pp2A (Lawson et al. 1990), and/or the phosphorylation of the RB and p53 proteins. It has been suggested (Lawson et al. 1990) that phosphorylation-dephosphorylation of T is a key mechanism in the regulation of cellular DNA synthesis mediated by T. Therefore, it is possible that other T functions, including its oncogenic activity, might be related to its phosphorylation state. Futhermore, the inactivation of RB and p53 cellular proteins by SV40 could require both their physical interaction with SV40 large T antigen, and an alteration of their phosphorylation state. This notion is supported by the observation that a low cellular T antigen concentrations, the addition of small t is essential for efficient anchorage-independent growth (Bikel et al. 1987). In contrast, at much higher T concentrations, a clear requirement for small t is not observed (Bikel et al. 1987).

IN VIVO STUDIES

SV40 oncogenicity

Wild-type SV40 is highly oncogenic in hamsters. Newborn animals are particularly subsceptible, usually developing fibrosarcomas at the injection site following subcutaneous (SC) inoculation of a low dose of SV40 (Eddy 1964). When newborn hamsters are inoculated with SV40 intracerebrally they develop ependymomas (Gerber and Kirschstein 1962). Wealing and adult animals may develop fibrosarcomas if injected SC with a high dose of virus, >10^9 plaque forming units (pfu), but only with a low tumor incidence and after prolonged incubation periods (Allison et al. 1967). When SV40 (>$10^{8.5}$ pfu) was injected intravenously into wealing hamsters, a circustance in which many cell types are exposed to high concentrations of the virus, lymphocytic leukemia, lymphoma, soft-tissue sarcoma and osteosarcoma developed (Diamandopoulus 1972). This diversity of tumors has not been reported by any other route of SV40 inoculation. These data suggest that in the SV40-hamster cell system, the type of cell rendered neoplastic by the virus, rather than the virus per se, determines the morphology of the induced neoplasm. These data also suggest that SV40 may be able to transform many different cell types, and that the type of cell transformed by SV40 is determined by the site of virus inoculation.

SV40 small t mutant oncogenicity

Studies to determine the oncogenicity of SV40 small t mutants indicate that these mutants are oncogenic when injected SC into newborn hamsters, but the tumors develop only after a prolonged latency (Dixon et al. 1982, Lewis and Martin 1979). The behavior of the tumors induced by one such deletion mutant, dl 884, differs from that of tumors induced by wild-type SV40 in another important way besides latency: A significant fraction of the hamster inoculated SC develop tumors distant from the SC site of injection (Dixon et al. 1982). To determine whether these altered tumorigenic properties arise from small t antigen lesions in general, or whether they pertain uniquely to the dl 884 mutant, we ewpanded this study (Carbone et al. 1989a, Matthews et al. 1987) to include additional mutants that are deleted in the small t antigen gene (Figure 1). Newborn random bred Syrian hamsters (in groups of 65 to 130 animals) were inoculated SC between the scapulae with 2 X 10^8 pfu of wt SV40; SV40 small t mutant dl 883, which has a 57 bp deletion spanning the small t donor splice site; dl 884, which contains a 247 bp deletion that also spans the small t donor splice site; or dl 890, which has a 27 bp deletion near the middle of the small t specific sequences. As noted in previous studies, the SC fibrosarcomas induced by the deletion mutants developed much more slowly than virus-induced tumors. Whereas the mean latency of wt virus-induced tumors was 35 weeks, for the deletion mutant-induced tumors the mean latency was 59 weeks. We also observed virus-specific differences in the site of tumor appearance. Fibrosarcomas at the site of virus inoculation developed in both wt SV40 and SV40 small t mutant injected animals. However, in addition to fibrosarcomas, hamsters injected with SV40 small t mutants dl 883, dl 884, or dl 890 developed abdominal tumors. The latency of abdominal tumor appearance was 34 weeks, considerably shorter than the latency of mutant induced fibrosarcomas (59 weeks), but similar to the latency of wt-induced fibrosarsomas (35 weeks).

Characterization of SV40 small t mutant-induced tumors

In an attempt to understand the unusual tumorigenic properties of the SV40 small t deletion mutants, we studied 12 cell lines obtained from different abdominal tumors induced by SV40 small t mutants dl 883, dl 884, or dl 890 (Carbone et al. 1989a). All of these cell lines were shown to be transformed by SV40. Almost 100% of the cell of each line contained nuclear SV40 T-antigen as detected by immunofluorescence. In addition all of these cell lines expressed the SV40 96 kDa large T antigen, but we could not detect the 17 kDa small t-antigen. Four of the cell lines were also tested for the presence of integrated SV40 viral DNA by Southern blot hybridization of restriction nuclease digested cellular DNA; all four lines contained viral DNA. Morphological and functional characterization of these lines indicated that they were derived from cells of the mononuclear phagocytic series, and they were defined as "true histiocytic lymphomas according to the human terminology. This conclusion was based on the following characteristics of these cell lines: 1) They generally adhered to glass and plastic; 2) they expressed class II MHC antigens; 3)

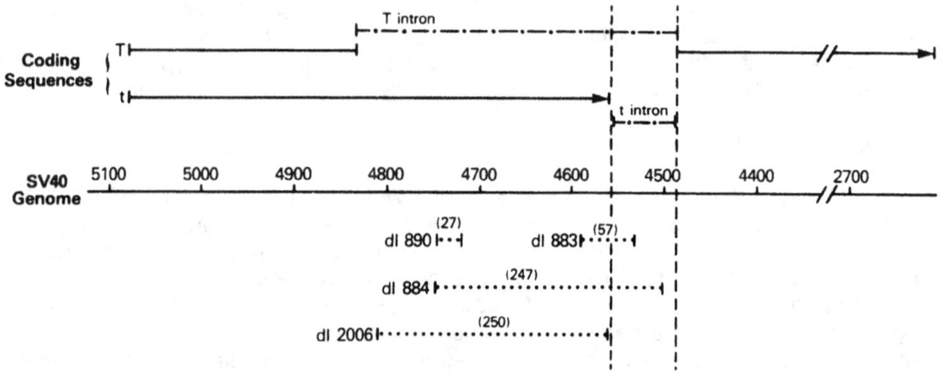

FIG.1 Map of the SV40 early region. The sizes (in parentheses) and positions of the small t antigen deletions are indicated below the genome map.

they were all positive for the nonspecific esterase reaction; 4) they produced lysozyme and fibronectin; 5) they actively phagocytosed; and 6) they expressed the MAC-2 antigen which is specific for a unique subpopulation of mononuclear phagocytes. Thus, the small t deletions in the SV40 genome appeared to permit the virus to transform cells that are distant from the site of virus inoculation; at this distant from the site of virus inoculation; at this distant site, the cells transformed are of a specific lineage, MAC-2+ peritoneal macrophages. Interestingly, Choi et al. (1988), reported that transgenic mice expressing the SV40 large T and small t antigens, under the transcriptional control of the mouse mammary tumor virus (MMTV) promoter, developed lymphomas and adenocarcinomas, but only lymphomas developed in transgenic mice exclusively expressing the SV40 large T and lacking the small t antigen. Together with our results, these experiments suggest that lack of small t expression may be associated with the preferential transformation of hematopoietic cells.

An additional small t mutant: dl 2006

We next decided to investigate the oncogenicity of an additional SV40 small t mutant, dl 2006, which contains a deletion of 250 bp that does not extend into the small t intron. We injected 41 random bred Syrian hamsters SC between the scapulae with 2 X 10^8 pfu of dl 2006 virus. We found that the oncogenicity of this mutant differed from the oncogenicity of the other mutants we had studied previously. When dl 2006 was injected SC into newborn animals, fibrosarcomas with a prolonged latency developed at the site.of inoculation; however, the tumours induced at a distant site were osteosarcomas rather than lymphomas. The cells derived from these tumours stained for nuclear SV40 antigen, and the 96 kDa T antigen was detected by immunoprecipitation.

We considered several possible explanations for the prolonged latencies and the altered spectrum of tumor types induced by the small t antigen deletion mutants. One interpretation might be related to the finding that the deletion mutant viruses are defective in the transformation of stationary cells (Martin et al. 1979, Sleigh et al. 1978). This finding has led to the suggestion that the small t protein may act as a growth factor. In tis absence, more time may be required for a tumor to develop to the stagez at which its growth is no longer restricted by internal cellular or external host factors. The anomaluos occurence of rapidly proliferating lymphomas and osteosarcomas in outbred hamsters injected SC with the small t deletion mutants might also be explained on the basis of a growth factor effect. In the absence of this putative small t product, the mutant viruses would lack the capacity to induce the proliferative state required for transformation, and might preferentially transform rapidly proliferating cells, e.g., lymphoblasts and osteoblasts, rather than resting cells such as fibroblasts. This explanation could account for the occurence of lymphomas and osteosarcomas with a relatively short latency (equal to that of wt SV40-induced SC fibrosarcomas), as well as the prolonged latencyt of SC fibrosarcomas in those animals that developed this tumor. We attempted to explore the possibility that small t act as a growth factor, but we did not detect any difference in the ability of media conditioned by wt SV40 or small t mutant transformed cells to support DNA replication in NRK indicator cells (Carbone et al. 1989b). These results suggest that cells transformed by wt SV40 and cells transformed by small t mutants secrete the same types of growth factors, but they do not entirely rule out the possibility that small t may act as a growth factor. In fact, small t might have a growth stimulatory effect within the cell, and/or once the transformation event has occurred this function could be dispensable.

The induction by SV40 of lymphomas and osteosarcomas is not without precedent. In experiments with wealing and adult hamsters, intravenous injection of a wild-type strain of SV40 efficiently induced lymphomas, leukemias and osteosarcomas (Diamandopoulus 1972). The latency of the tumors induced by the intravenous route was somewhat shorter (16-26 weeks) than the tumour latencies seen in our experiments using the SC route. That the wild-type virus is capable of inducing these types of tumors, when introduced directly into the blood stream, suggests that the wild-type virus is able to transform lymphoblasts and osteoblasts when they are available in high concentration. However, when inoculated SC the wild type virus is capable of transforming the slowly proliferation fibroblasts efficiently in situ. The relatively early development of SC fibrosarcomas might suppress the emergence of tumors elsewhere. Alternatively, it is possible that the lymphomas and osteosarcomas induced following intravenous injection of the wild-type virus (Diamandopoulus 1972) were induced by viral mutants present in the wild-type virus stock. Finally, it is also possible that the small t antigen may in some way influence local absorbtion of the virus, such that in absence of this antigen, the virus is better able to disseminate widely. This phenomenon might also be mediated by the host immune system; however, studies by us (Haddada et al. 1989) indicate that both wt SV40 and SV40 mutant-transformed cells express the same high levels of class I major histocompatibility antigens, confirming previous studies (Tevethia et al. 1980) in which small t deletion mutant viruses were shown to immunize mice, and to induce cytotoxic lymphocytes as well as transplantation antigene, with essentially the same efficiency as wild-type virus.

In summary, our results, as well as others, indicate that the small t antigen plays an important role in both transformation and oncogenicity by SV40. However, several questions must be answered to assign a specific function to this oncogene product. First, why do SV40 small t mutants induce tumors at a site remote from that of virus inoculation? To understand this aspect of small t function, we investigated the oncogenicity of SV40 small t mutants under conditions which allowed many different cell types to be exposed to high concentrations of the virus. We questioned whether the small t mutants dl 883, dl 884, and dl 890, injected SC, induce MAC-2+ macrophage lymphomas because they have aquired a specific cell tropism. If so, they should also induce lymphomas when injected directly into the bloodstream. On the other hand, if the appearance of lymphomas is not related to a cell tropism, but rather to other factors such as a local absorption of the virus or a response of the host immune system, a diversity of tumor types would be expected to develop after inoculation of the mutants into the bloodstream. We asked the same question for dl 2006, which when injected SC induced osteosarcomas rather than lymphomas at a distant site.

Intravascular inoculation of the small t mutants

In preliminary experiments, we injected 10^8 pfu of various SV40 viruses into the left cardiac ventricles of random-bred, 21 days old Syrian hamsters. Groups of 12 animals each were injected with SV40 small t mutant dl 883; dl 2006; or wild type SV40 830 (Table 1). In addition 12 animals were injected with medium alone, and 12 animals served as uninjected controls. Animals developing tumors were sacrificed when they appeared ill, and all surviving animals with or without tumors were killed and necropsied 9 months following injection. A detailed microscopic examination was undertaken of the viscera, brain, and bones of all animals in this study. Tumor cell lines were also established.

The results of the intracardiac (IC) injection experiments are summarized in Table 1. One to three hamsters in each group died a result of the IC injection. Of the surviving animals inn the group injected with dl 883, all developed lymphomas. The group injected with dl 2006 developed various tumors: mainly lymphoma (55%), but also osteosarcoma, adenocarcinoma and mesothelioma. One hamster in this group developed both an osteosarcoma and a lymphoma, and two hamsters did not developed tumors. The group injected with the wild type virus als developed a variety of tumors: However in this group, several of the tumors were pleural mesotheliomas (56%). In addition, lymphoma, osteosarcoma and the unjniected controls did not develop any tumors in the 9 month observation period.

The mean latency of tumor development (Table 2) was 31 weeks for dl 2006 induced tumors, 29 weeks for dl 883 induced tumors, and 23 weeks for wt SV40 induced tumors. The mean latency was correlated with the induciong virus rather than with a particular tumor type: For example, lymphomas induced by wild type 830 had a mean latency of 20 weeks, while lymphomas induced by dl 2006 had a mean latency of 30 weeks (Table 2). No differences in latency of tumor types were apparent between males and females, with the exception of mesotheliomas which seemed more likely to develop in females (5 out of 6).

The various tumor types were identified by their macroscopic appearance and microscopic characteristics. Lymphomas were further characterized to determine the transformed cell type (e.g., mononuclear phagocyte, B or T lymphocyte, granulocyte). Cytochemical staining for a variety of enzymes that distinguish these different cell types (Carbone et al. 1989a), the phagocit activity of the tumor cells, and fluorescence staining fro surface immunoglobulins (specific for B-cells) and Mac-2 antigen (specific for histiocytes), indicated that the hamsters in this study developed two different types of lymphoma: Lymphocytic lymphoma and histiocytic lymphoma. Some animals developed both types of lymphoma but we do not know whether this is the result of two different transformation events, or whether the parent transformed cell differentiated into two different tumor types.

Fresh tumor material was used to establish cultures of the various tumor types. All of the cell lines established were shown to be transformed by wt SV40 or SV40 mutants. Immunoprecipitation experiments using antisera from hamsters bearing SV40 tumors indicated that all of the cell lines expressed 96 kDa SV40 large T antigen, but only the cell lines derived from wt SV40-induced tumors expressed the 17 kDa SV40 small t antigen (Figure 2). Interestingly, a truncated form of the small t antigen was detected in cell lines derived from tumors induced by dl 2006 (Figure 2). To confirm that these cell lines were representativeof the primary tumor, DNA was extracted from frozen tumor specimens and from the corresponding cell lines, run through an agarose gel, blotted and hybridized with ^{32}P- labeled SV40 DNA according to standard procedures (Southern 1975). The integration site of SV40 DNA was the same in the primary tumors as in the established cell lines.

Table 1 - Various tumor types (percentage) developing in hamsters injected with SV40 viruses (see text). Groups of 12 animals each were injected IC with SV40 viruses.
3 animals in the wt and dl 883 groups, and 1 animal in the dl 2006 group died as a result of the IC injection and they are not considered in the results reported in this table. One of the animals in the dl 2006 group developed 2 different tumors.

Tumor type	SV40 virus		
	wt 830	dl 883	dl 2006
lymphoma	22%	100%	55%
mesothelioma	56%	0%	9%
osteosarcoma	11%	0%	18%
myxoma	11%	0%	0%
adenocarcinoma	0%	0%	9%
animals not developing tumor	0%	0%	18%

Table 2 - Tumor latency associated with various SV40 viruses. Mean latency defined as age (in weeks) at which hamsters were sacrificed following tumor development.

tumor type	virus		
	wt 830	dl 883	dl 2006
all tumors	23	29	31
lymphoma	20	29	30
mesothelioma	24	-	33
osteosarcoma	26	-	31
myxoma	20	-	-
adenocarcinoma	-	-	35

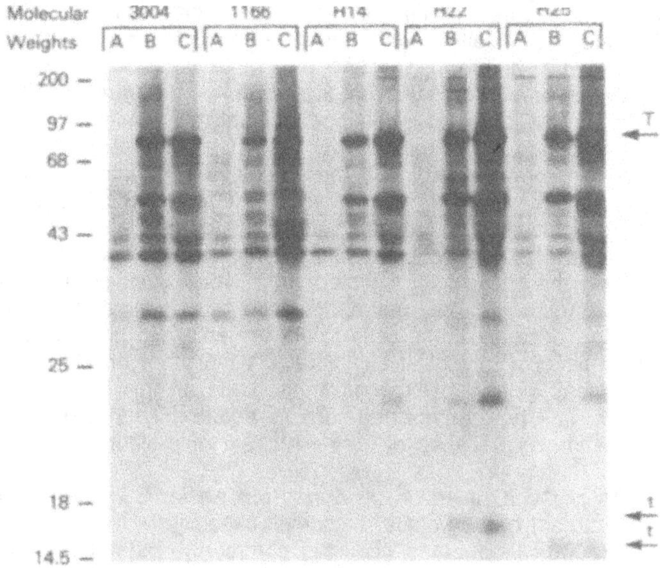

FIG.2 Immunoprecipitation of SV40 T antigens. The large T and the small t antigens migrate at the expected mol. wt. of 96 kDa and 17 kDa respectively. The cell lines studied are: 3004 (dl 883-induced lymphoma derived cell line); 1166 (wt SV40-induced fibrosarcoma derived cell line); H14 (dl 883-induced lymphoma derived cell line); H22 (wt SV40-induced mesothelioma derived cell line) and H25 (dl 2006-induced mesothelioma derived cell line).
Lanes A, normal hamster serum, lanes B, mouse monoclonal antibody anti-T and t SV40 antigens, (Oncogene Science, Manhasset, NY 11030 USA), lanes C, serum derived from SV40 hamster bearing tumors.

DISCUSSION

These preliminary results of the IC injection of SV40 small t mutants offer further support to the notion, that these mutants have a markedly reduced ability to transform non- proliferating cells (Martin et al. 1979; Seif and Martin 1979; Sleigh et al. 1978). The small t mutants are highly oncogenic when injected IC into random bred Syrian hamsters. However, significant differences in oncogenicity were observed between wt SV40 and the mutants and between the two different mùutans injected. Tumors induced by small t mutants developed with a prolonged latency, vis-a'-vis wt SV40, regardless of their histologic origin. The SV40 small t mutant dl 883 yielded only hematopoietic tumors, even though a diversity of cell types was exposed to high concentrations of the virus as a consequence of intravascular inoculation. These results confirmed as previous experiments (utilizing the SC injection route) suggesting that the oncogenicity of the dl 883 mutant is restricted to hematopoietic cells (Carbone et al. 1989a). However in the earlier experiments, only a specific subpopulation of hematopoietic cells (MAC-2+ peritoneal macrophages) was transformed. Those results led us to postulate that the mutated viruses had acquired a specific tropism for this unique cell type.

The results presented in this report indicate that while dl 883 preferentially transforms hematopoietic cells, more than one cell type can be transformed within this class of cells (i.e. macrophages and B-lymphocytes). Alternatively, dl 883 may transform the same precursor cell type that further differentiates into macrophages or B-lymphocytes. The latter possibility seems unlikely since no B-cell lymphomas were observed following SC injection of dl 883 (Carbone et al. 1989a). Rather we suggest that to transform cells, dl 883 must achieve a certain intracellular threshold of viral partilce concentration. When the virus is injected SC this threshold level is reached at the site of injection, resulting in fibrosarcomas with a prolonged latency, but it is also reached in mononuclear phagocytes that normally are the first line of defense in clearing the body of the injected virus (Abramczuk et al. 1984). Among the various macrophage lineages only a subpopulation of peritoneal MAC-2+ macrophages will be transformed because these cells retain the ability to proliferate despite a high degree of differentiation (Stewart et al. 1980). On the other hand, when the animal is injected IC, most cell types are exposed to the virus. In this istance, MAC-2 positive macrophages will still be transformed, but in addition lymphoid cells will also be transformed, presumably because of their high rate of proliferation.

The results of the IC injection confirmed our earlier SC experiments suggesting that the dl 2006 mutant has a unique oncogenic capacity. In fact, among the hamsters injected with dl 2006, lymphomas occurred commonly, but other tumor types were also induced (Table 1). Thus the oncogenicity of dl 2006, when injected IC, is different than that of either dl 883 or wt SV40: Lymphomas are favored, unlike wt, but other tumors occur as well, unlike dl 883. Interestingly, while dl 883 cannot produce any form of small t antigen because of the location of its deletion, which comprises the donor splice (Figures 1 and 2), dl 2006 has a 250 bp deletion that does not prevent the production of a truncated form of small t antigen (Figures 1 and 2). This truncated form of small t may account for a partial complementation of the large T antigen. In other words, the putative ability of small t to induce the resting cell to traverse from G1 to S might be retained by this truncated product, albeit with less efficiency than the intact small t antigen (explaining the prolonged tumor latency but greater diversity of tumor types associated with dl 2006). The biochemical function of the small t antigen is still unknown: The possibility that different domains of this oncoprotein exert different functions is particularly attractive in light of recent experiments which demonstrate that the small t antigen binds to the regulatory and the catalytic subunits of phosphatase 2A. Is the truncated version of small t produced by dl 2006 able to associate with this enzyme? If so, the small t-pp2A complex may play a critical role in SV40 oncogenicity, influencing the cell type that becomes transformed; on the other hand, if the small t antigen produced by dl 2006 fails to associate with phosphatase 2A, the putative t-pp2A complex may relate to latency but not to cell tropism. In either case, studies of the truncated version of the small t antigen produced by dl 2006 should shed further light on the biology of the small t oncogene.

Finally, it is of considerable interest that hamsters, injected IC with wt SV40, preferentially developed mesotheliomas. Mesothelial cells are phagocytic cells: Therefore they probably phagocytose virus particles and this may account fot the mesotheliomas we observed. Despite the fact that mesotheliomas in humans have never been associated with a virus etiology, SV40-induced mesothelioma in the hamster might be a useful model for understanding the molecular mechanism that lead to mesothelioma in man. In particular, the known associations between the large T antigen of SV40 and the products of the tumor suppressor genes RB and p53 point to alterations of these cellular oncogenes as possible events in human mesothelial cell transformation.

SUMMARY

The large T antigen, encoded by an SV40 oncogene, induces DNA replication in infected cells and cellular transformation. The small t antigen, created by differential splicing of the same transcript, facilitates T antigen-induced DNA replication and transformation in resting cells, but it is not needed in dividing cells. We have found that while subcutaneous injection of newborn hamsters with wild type SV40 induces in situ fibrosarcomas, small t deletion mutants of SV40 more readily induce tumors of rapidly dividing cells, e.g., lymphoblasts and osteoblasts, with the latter tumors appearing distal to the site of injection. When wild type SV40 is injected into the cardiac ventricles, a diversity of tumor types results, with a preponderance of pleural mesotheliomas. However, the intracardiac injection of the small t mutants mainly yelds lymphomas and osteosarcomas. These in vivo results are consistent with the earlier in vitro results, and lend further support to the notion that small t is needed for the induction of DNA replication and transformation of non-proliferating cells by SV40. The basis of this requirement may lie in the known association of t with pp2A: The t-pp2A complex may affect the phosphorylation state of T antigen, as well as its capacity to induce DNA replication. Alternatively, this complex may affect the ability of T to inhibit RB and p53, anti-oncogene products which suppress DNA replication.

BIBLIOGRAPHY

1. Abramczuk, J., Pan, S., Maul, G., and Knowles, B.B. 1984. Tumor induction by Simian Virus 40 in mice is controlled by long-term persistence of the viral genome and the immune response of the host. J. Virol. 49, 540-548.
2. Allison, A. C., Chesterman, F.C., Baron, S. 1967. Induction of tumors in adult hamsters with simian virus 40. J. Natl. Cancer Inst. 38, 567-527.
3. Bikel, I., H., Hamon, E. L., Brown, J., Boltax, M., McCormak, J., Boltax, and D. Livingston. 1987. SV40 small t antigen enhances the transformation activity of limiting concentrations of SV40 large T antigen Cell 48, 321-330.
4. Bischoff, J.R., Friedman, P.N., Marshak, D.R., Prives, C., and Beach, D. 1990. Human p53 phosphorylated by p60-cdc and cyclin B-cdc2. Proc. Natl. Acad. Sci. USA. 87, 4766-4770.
5. Butel, J.S. 1986. SV40 large T-antigen: dual oncogene. Cancer Surveys. 5, 343-365.
6. Carbone, M., and Levine, A.S. 1990. Oncogenes, antioncogenes and the regulation of cell growth. Trends Endo. Metab. 1; 248-253.
7. Carbone, M., Lewis, A.M., Matthews, B., Levine, A.S., Dixon, K., 1989a. Characterization of hamster tumors induced by simian virus 40 small t deletion mutant as true histiocytic lymphomas. Canc. Res. 49, 1565-1571.
8. Carbone, M., Kajiwara, E., Patch, C., T., Lewis, A.M., Jr, Levine, A.S., and Dixon, K., 1989-b. Biochemical properties of media conditioned by simian virus 40-induced hamster tumor cells: correlation with distinct cell phenotypes but not with oncogenicity. Cancer Res. 49, 6809-6812.
9. Choi, Y., Lee, I. and Ross, S.R., 1988. Requirement for the simian virus 40 small tumor antigen in tumorigenesis in transgenic mice. Mol. Cell. Biol. 48, 3382-3390.
10. Cooper, A.F., and Whyte, P., 1989. RB and the cell cycle: entrance or exit? Cell, 58, 1009-1011.
11. Diamandopoulus, G.T., 1972. Leukemia, lymphoma, and osteosarcoma induced in the Syrian golden hamster by simian virus 40. Science 176, 173-175.
12. Dixon, K., Ryder, B.J., and Burch-Jaffe, E., 1982. Enhanced metastatis of tumors induced by a SV40 small t deletion mutant. Nature (Lond.), 296, 672-675.
13. Eddy, B., E., Borman, G.S., Grubbs, G.E., and Young, R.D., 1964. Identification of the oncogenic substance in greeen monkey kidney cell cultures as simian virus 40. Virology 17, 65-75.
14. Fearon, R.E., and Vogelstein, B., 1990. A genetic model for colorectal tumorigenesis. Cell 61, 759-767.
15. Gerber, P., and Kirschstein, R.L., 1962. SV40-induce ependymomas in newborn hamsters. Virology 18, 582-588.

16. Green, M. R., 1989. When the products of oncogenes and antioncogenes meet. Cell 56, 1-3.
17. Haddada, H., Sogn, J.A., Coligan, J.E., Carbone, M., Dixon, K., Levine, A.S., and Lewis, A.M., Jr. 1988. Viral gene inhibition of class I major histocompatibility antigen expression: not a general mechanism governing the tumorigenicity of adenovirus type2-, adenovirus type 12-, and simian virus 40- transformed syrian hamster cells. J. Virol. 62, 2755-2761.
18. Lane, D.P., and Benchimol, S., 1990. p53: Oncogene or Antioncogene? Genes & Development 4, 1-8.
19. Lawson, R., Cohen, P., and Lane, D.P. 1990. Simian virus 40 large T-antigen-dependent DNA replication is activated by protein phosphatase 2A in vitro. J. Virol. 64, 2380-2383.
20. Lewis, A.M., Jr., and Martin, R.G. 1979. Oncogenicity of Simian virus 40 deletion mutants that induce altered 17-kilodalton t-proteins. Proc. Nat. Acad. Sci. USA. 76, 4299-4302.
21. Martin, R.G., Setlow, V.P., Edward, C.A.F., and Vembu, D., 1979. The roles of the simian virus 40 tumor antigens in transformation of Chinese hamster lung cells. Cell 17, 635-643.
22. Matthews, J.B., Levine, A.S., and Dixon, K. 1987: Deletion mutations in the small t antigen gene alter the tissues specificity of tumors induced by simian virus 40. Virology. 61, 1282-1285.
23. Pallas, C.D., Shahrik, L.D., Martin, L.B., Jaspers, S., Miller, B.T., Brautigan, L.D., and Roberts, M.T. 1990. Polyoma small and middle T antigens and SV40 small t antigen form stable complexes with protein phosphatase 2A. Cell 60, 167-176.
24. Rundell, D., and J., Cox, 1979. Simian virus 40 t antigen affects the sensibility of cellular DNA synthesis to theophylline. J. Virol. 30, 394-396.
25. Sleigh, M.J., Topp, W.C., Hanich, R., and Sambrook, J.F., 1978. Mutants of SV40 with an altered t small t protein are reduced in their ability to transform cells. Cell 14, 79-88.
26. Southern, E.M., 1975. Detection of specific sequences among DNA fragments separated by gel electrophoresis. J. Mol. Biol. 98: 403-517.
27. Stewart, C.C. 1980. Formation of colonies by mononuclear phagocytes outside the bone marow. In: R. van Furth (ed.), Mononuclear Phagocytes: Functional Aspects, pp. 377-413. The Haque: Martinus Nijoff Publishers.
28. Seif, R., and Martin, R.G., 1979. Simian virus 40 small t antigen is not required for the manteinance of transformation but may act as a promoter (cocarcinogen) during establishment of transformation in resting rat cells. J. Virol. 32, 979-988.
29. Tevethia, S. S., Flyer, D.C., and Topp, W.C., 1980. Biology of simian virus 40 (SV40) transplantation antigen (TrAg). VII. Induction of SV40 TrAg in nonpermissive mouse cells by early viable SV40 deletion (0.54/0.59) mutants. Virology 195, 488-496.
30. Topp, W.C., Lane, D., and Pollack, R., 1981. Transformation by SV40 and polyoma virus. In: J. Tooze (ed.), DNA Tumor Viruses, Ed. 2, Part 2, revised, pp. 205-296. Cold Spring Harbor, NY: Cold Spring Harbor Laboratory.
31. Walter, G., Ruediger, R., Slaughter, C., and Mumby, M. 1990. Association of Protein phosphatase 2A with polyoma virus middle tumor antigen. Proc. Natl. Acad. Sci. USA 87, 2521-2525.
32. Weinberg, R.A. 1989. Oncogenes, antioncogenes, and the molecular basis of multislep carcinogenesis. Cancer Res. 49, 3713-3721.

MOLECULAR BASIS OF HUMAN CANCER
May 21 - June 1, 1990
Erice, Sicily, Italy

PARTICIPANTS

Stuart AARONSON
Lab. of Cell. & Mol. Biology
NCI, NIH
Bldg 37 IE-24
20205 BETHESDA, USA

Francesco ANTOLINI
Dip. di Biochimica Cellulare
Università di Perugia
Via del Giochetto, 2
06100 PERUGIA, Italy

Asok ANTONY
Dept. of Medicine-Hematology
Indiana University
975 West Walnut Street
INDIANAPOLIS, IN 46202-5121, USA

Ilia ARANDA
University of Mexico City
MEXICO

Huseyin BAGCI
Dept. Medical Biology Genetics
Faculty of Medicine
Akdeniz University
07059 ANTALYA, Turkey

Mariano BARBACID
Dept. of Molecular Biology
Squibb Inst. Medical Research
P.O. Box 4000
PRINCETON, NJ 08543, USA

Nursen BASARAN
Faculty of Pharmacy
Hacettepe University
Pharmaceutical Toxicology
06100 ANKARA, Turkey

Mandy BOHN
Dept. of Microbiology
Loma Linda Unversity
1101 Anderson
LOMA LINDA, CA 92350, USA

Andrew BRADBURY
CNR Ist Neurobiologia
Via Carlo Marx, 5
00156 ROMA, Italy
MRC Lab of Molecular Biology
Hills Road
CAMBRIDGE CB2 2QM, UK

Antonia BRADBURY
Dept. of Biological Chemistry
University of California
School of Medicine
DAVIS, CA 95616, USA

Metin CAKMAKCI
Dept. of Surgery
Hacettepe University Fac. Med.
Hacettepe
06100 ANKARA, Turkey

Michele CARBONE
NICHD-NIH
Bldg. 6 Rm 140
BETHESDA, MD 2089, 2USA

Paolo CATASTI
Istituto di Biofisica
Facoltà di Medicina
Università di Genova
Via Giotto, 2
16100 GENOVA, Italy

Webster CAVANEE
Ludwig Institute
687 Avenue des Pines Ovest
H3A 1A1 MONTREAL, Canada

Maurizio CIANFRIGLIA
Lab. di Immunologia
Istituto Superiore di Sanità
Viale Regina Elena, 299
00161 ROMA, Italy

Mehmet COLAKOGLU
Dept. Obstetrics &Ginecology
P.O. Box 103 Nalcalcl.
42005 KONYA, Turkey

Carlo CROCE
Fels Research Institute
Temple University
Medical Research Building
3420 N. Broad St.
PHILADELPHIA, PA 19103, USA

Nejat DALAY
Dept. of Basic Oncology
Oncology Institute
Capa
3439§0 ISTANBUL, Turkey

Ramazan DEMIR
Dept. Histology & Embryology
Faculty of Medicine
Akdiniz University
07059 ANTALYA, Turkey

Emilio DI MARIA
Istituto di Biofisica
Facoltà di Medicina
Università di Genova
Via Giotto, 2
16153 GENOVA, Italy

Carla EMILIANI
Dip. di Biochimica Cellulare
Università di Perugia
Via del Giochetto
06100 PERUGIA, Italy

Raymond ERIKSON
Dept. of Cellular & Dev. Biol.
Harvard University
16 Divinity Ave.
02139 CAMBRIDGE, USA

Ethel Michele DE VELLEARS
German Cancer Res. Center
Im Neunheimer Feid 280
D-6900 HEIDELBERG 1, FRG

Paola GAVAZZO
Istituto di Biofisica
Facoltà di Medicina
Università di Genova
Via Giotto, 2
16153 GENOVA, Italy

Eugene GRIVSKY
Dept. of Organic Chemistry
The Welcome Research Labs.
4407 Eastwood Court
22032 FAIRFAX, USA

Theodor KARALIS
Democratis University of Thrace
67100 XANTHI, Greece

Irit LAX
Dept. of Molecular Biology
Roner Biotechnology
680 Allendae Rd.
KING OF PRUSSIA, PA 19066, USA

Jane LIAN
Dept. of Cell Biology
University of Mass. Med. School
55 Lake Ave. North
WORCESTER, MA 01655, USA

Guven LULECI
Dept. of Medical Biology
Faculty of Medicine
Akdeniz University
07059 ANTALYA, Turkey

Werner LUTZ
Dept. of Toxicology
University of Zurich
Schorenstrasse 16
CH-8603 SCHWERZENBACH, Switzerland

Giancarlo MASCETTI
Istituto di Biofisica
Facoltà di Medicina
Università di Genova
Via Giotto, 2
16153 GENOVA, Italy

Giulio MILAZZO
Società Bioelettrochimica
Piazza G. Verdi, 9
00198 ROMA, Italy

Claudio NICOLINI
Istituto di Biofisica
Facoltà di Medicina
Università di Genova
Sal. Sup. Noce, 35
16131 GENOVA, Italy

Mario NIZZARI
Istituto di Biofisica
Facoltà di Medicina
Università di Genova
Via Giotto, 2
16153 GENOVA, Italy

Jennifer PIETENPOL
Dept. of Cell. Biology
Vanderbilt University
C-2310 MCN
NASHVILLE, TN 37232, USA

R. PRASAD
Dept. of Pathology
University of Illinois
1853 W. Polk St.
60466 CHICAGO, USA

Simonetta PULCIANI
Dip. di Virologia
Ist. Superiore di Sanità
Viale Regina Elena, 299
00161 ROMA, Italy

Kevin RANEY
Dept. of Chemistry
Vanderbitt University
Station B
NASHVILLE, TN 37235, 2USA

Nils RINGERTZ
Dept. of Medical Cell Genetics
Medical Nobel Institute
Karolinska· Inst. Box 60400
S-10401 STOCKHOLM, Sweden

Yoseph SCHLESSINGER
Pharmacology Department
New York University
550 First Avenue, Room MSB 424
NEW YORK, NY 10016, USA

Steven SMOCK
Dept. of Cell. Biology
University of Massachusetts
55 Lake Ave. North
WORCESTER, MA 01655, USA

Giuseppe STARACE
CNR
Istituto di Medicina Sperimentale
Viale Marx, 15
00137 ROMA, Italy

Gary STEIN
Dept. of Cell. Biology
University of Mass.
55 Lake Avenue North
WORCESTER, MA 01655, USA

Henkie TAN
Dept. of Surgery & Microbiol.
Loma Linda University
1101 Anderson St.
LOMA LINDA, CA 92350, USA

Orhan TERZIOGLU
Dept. Medical Biology Genetics
Dokuz Eylul Univ. Fak. Med.
Inciralti
35340 IZMIR, Turkey

Belma TURAN
Dept. of Biophysics
Ankara University Med. Fac.
Sihhiye
06100 ANKARA, Turkey

George WANDE WOUDE
National Center Institute
Frederick Cancer Res. Faculty
P.O. Box B
FREDERICK, MD 21701, USA

Laura VERGANI
Istituto di Biofisica
Facoltà di Medicina
Università di Genova
Via Giotto, 2
16153 GENOVA, Italy

Peter VOGT
MILRO
USL
2011 Zonal Ave. HMR 401
90033 LOS ANGELES, USA

Peter YAO
Dept. Med. Biolog. Chemistry
University of California
School of Medicine
DAVIS, CA 95616, USA

Luo YIN
Dip. di Biologia Molecolare
Istituto Nazionale Ricerca sul Cancro
Viale Benedetto XV, 10
16132 GENOVA, Italy
Dept. of Cell. Biology
Cancer Institute
Chinese Academy of Medical Sciences
BEIJING 100021
PR China

Zhen ZHU
Dip. di Cancerogenesi Chimica
Ist. Nazionale Ricerca sul Cancro
Viale Benedetto XV, 10
16132 GENOVA, Italy
Dept. of Cell Biology
Cancer Institute
Chinese Academy of Medical Sciences
BEIJING 100021
PR China

Harald ZUR HAUSEN
German Cancer Res. Center
Im Neuenheimer Feid 280
D-6900 HEIDELBERG 1, FRG

INDEX

www.ingramcontent.com/pod-product-compliance
Lightning Source LLC
LaVergne TN
LVHW052314200926
848481LV00002B/563

* 9 7 8 1 4 8 9 9 2 5 6 5 7 *